Towards Excellence

Leading a Doctoral Mathematics Department in the 21st Century

Towards Excellence

Leading a Doctoral Mathematics Department in the 21st Century

AMERICAN MATHEMATICAL SOCIETY
TASK FORCE ON EXCELLENCE

John Ewing, Editor

Cover art courtesy of and reproduced with permission of Coqui Calderón: *The Way Through Darkness*, acrylic on canvas, 48″ × 60″, 1991. The original painting is owned by the American Mathematical Society.

2000 *Mathematics Subject Classification*. Primary 00A20.

Library of Congress Cataloging-in-Publication Data

Towards excellence : leading a doctoral mathematics department in the 21st century / John Ewing, editor ; American Mathematical Society Task Force on Excellence.
p. cm.
ISBN 0-8218-2033-8 (alk. paper)
1. Ewing, John, 1944– . II. American Mathematical Society. Task Force on Excellence.
QA13.T68 1999
510′.71′173–dc21 99-30371
CIP

This material is based upon work supported by the National Science Foundation under Grant RED-9550471 and by a grant from the Exxon Education Foundation. Any opinions, findings, and conclusions or recommendations expressed in this material are those of the authors and do not necessarily reflect those of the National Science Foundation or the Exxon Education Foundation.

10 9 8 7 6 5 4 3 2 06 05 04 03 02 01 00

Acknowledgments

The American Mathematical Society

and

The Task Force on Excellence

gratefully acknowledge the support for this project by

The National Science Foundation

under Grant RED-9550471

and

EXXON Education Foundation

under Grants 03/1994 and 04/1995

Disclaimer

Any opinions, findings, conclusions, or recommendations expressed in this publication are those of the authors and do not necessarily reflect the view of the organizations that provided financial support for this project.

Headquarters of the AMS, Providence, RI

Task Force on Excellence
American Mathematical Society
www.ams.org

For an electronic copy of this publication visit
http://www.ams.org/towardsexcellence

On the Cover

The Way Through Darkness, **by Coqui Calderón**
acrylic on canvas 48" x 60", 1991

Cover art courtesy of and produced with permission of Coqui Calderón
Original painting is owned by the American Mathematical Society,
on display in the Hille Conference Room
Providence, RI

Contents

Part V: Resources

Appendices

Foreword
Who Wrote This Book?

In many ways this is a book written by a committee. Every member of the Task Force on Excellence (see Appendix A) participated in focus groups and committee discussions; every member read and critiqued all the material; every member contributed to parts of the actual writing. The Task Force decided to make most of that writing anonymous, however, to emphasize that this was a collective effort, representing the experience of not only the Task Force but more than one hundred faculty, chairs, and deans.

Nonetheless, there was one person whose effort was extraordinary and who contributed the heart of this book, Part I: Conclusions. Jim Lewis of the University of Nebraska spent many weeks and months writing and rewriting those first four chapters. While they represent the collective view of the entire Task Force, that view was shaped and focused by Jim's experience and wisdom. He wrote those four chapters. He not only wrote but he also listened, accepting both praise and criticism with remarkable grace. We are all grateful for his uncommon effort; this book would not exist without Jim Lewis.

Finally, none of this work would have been possible without the leadership over the last five years of Mort Lowengrub, Dean of the College of Arts and Sciences at Indiana University and Chair of the Task Force. His many presentations at meetings and focus groups shaped the final form of this book. His enthusiasm and vision kept the book on target and made its purpose to help research mathematicians, not to criticize them. His faith in the value of mathematics shows throughout this book.

John Ewing

Preface

This publication was written by and for mathematicians who work in America's research universities. It is aimed at faculty who work in mathematics departments[1] granting Ph.D.s. We hope the material is useful to other faculty (for example, in departments of statistics or in liberal arts colleges), but we are speaking most directly to mathematicians in research universities.

The idea that led to this publication is simple. The American mathematics departments awarding doctoral degrees produce most of our future mathematicians and much of our mathematics research. If we want American mathematics to be healthy, these departments must be healthy as well. We need to give them a prescription for health — a recipe for creating an excellent department that not only deserves but also secures from its university the necessary resources for excellence.

When the Task Force on Excellence began its work, the approach was simply: "How do we make the case to the dean for more resources?" But that approach assumed that all departments received inadequate resources, had similar needs, and merited a greater share of a university's base. Of course, it is hard to argue that all mathematics departments are inadequately funded in comparison with their peers. All departments do not all have the same needs. And it is not possible to provide a prescription (at least publicly) for convincing every dean to move funds from other departments into mathematics. It soon became clear that the simple idea (a recipe for excellence) was illusory.

Eventually, the task force was drawn to a more fundamental idea: Mathematics departments should position themselves to receive new or reallocated resources by meeting the needs of their institutions. That does not mean sacrificing the intellectual integrity of an academic program, nor does it mean relegating mathematics to a mere service role. It *does* mean fulfilling a bargain with the institution in which one lives, and for most departments a major part of that bargain involves instruction.

The focus of the task force became finding ways in which research departments can enhance their instructional program, at both the undergraduate and graduate levels. If departments carry out this part of their mission (and for many departments, add outreach activities as well) in a way that brings credit to the department and distinction to the university, then the necessary resources for a healthy department should follow, at least consistent with the ability of each institution to support its academic programs. Benefits will accrue both to the department's instructional program *and* to its research program.

---

[1] There are 177 mathematics departments that award the doctoral degree in mathematics and comprise Groups I, II and III in the Annual AMS-IMS-MAA Report. (See Appendix A.) Most (about 70 percent) use the name, Department of Mathematics; Twenty-one call themselves Department of Mathematical Sciences; and 18 call themselves Department of Mathematics and Statistics. A variety of other names are also used. In this publication we will consistently refer to each department as the Department of Mathematics.

This is a simple idea that many people find either ridiculously obvious or insidiously subversive. Much of the material in this book is aimed at convincing the reader that it is neither. By describing examples, we show departments that they can find creative ways to position themselves better in their own institutions. This is not an obvious process. By sharing comments and views from many different people in many different departments, we hope to convince the reader that meeting the needs of one's institution does not subvert the fundamental mission of a research department, but rather makes a healthy research department possible.

Part I contains the background and conclusions of the task force. While putting the conclusions at the beginning may seem unusual, we believe the central message of this work should be stated clearly and immediately, in advance of the evidence. We also include with the background some commentary on what the Task Force could *not* accomplish in its work, as well as some cautions about the scope of this project.

Part II provides excerpts from the fourteen focus group discussions carried out by the Task Force, along with commentary that summarizes the messages from the various groups. These focus groups formed the basis for much of the Task Force's work. They provided an opportunity for groups of chairs, deans, and (in one case) young mathematicians to share concerns with the committee and with one another. While it is impossible to capture on paper the full exchange of views, the excerpts provide a glimpse of both the shared concerns and the individual successes of some departments.

Part III contains examples for departments to consider. These examples illustrate some ways in which departments can meet the needs of their institutions, and while they are not examples that all departments can emulate, they suggest the breadth of possibilities. The Task Force conducted five site visits, and each site was selected in order to understand a specific program or aspect of that department. A collection of shorter reports on other programs of interest is included as well.

Part IV includes some short essays that examine the ways in which the differing views of departments contrast and agree with each other. One fact became clear early in our work: Mathematics departments do not view themselves as others view them. Which view is correct? All are . . . and none are. It is essential, however, for mathematics departments to understand how others view them, and these essays are intended to begin the process of understanding.

Part V contains resources — material that departments might use for self-study or external reviews, as well as a list of books and articles that refer to many of the topics considered here.

Early in its work, the Task Force on Excellence was able to begin its work with grants from the Exxon Education Foundation. Those grants and the support of Bob Witte of Exxon were crucial to this work. Later, a substantial grant from the National Science Foundation allowed the Task Force to expand its focus and to reach a broader group. We are grateful to both the Exxon Foundation and the National Science Foundation for their support and continued encouragement throughout this project.

Throughout its work the Task Force has been supported by Raquel Storti from the American Mathematical Society. Her belief in this project, her dedication, and her enthusiasm made this project (and this book) possible. We are grateful for all that she has done over these past six years.

We stress that this small book was prepared by friends of mathematics, many of whom have had experience as a chair, dean, provost, or even president of a research university. If some comments are perceived as criticism, please accept them as criticism from a group of mathematicians who have spent their careers among research mathematicians, and who are thankful for the opportunity.

The Task Force on Excellence

Part I

Conclusions

Chapter 1
Background

We have a simple message: To ensure their institution's commitment to excellence in mathematics research, doctoral departments must pursue excellence in their instructional programs.

Most reports about resources for mathematics research have focused on federal funding. This book is different in that it focuses on the health of universities and especially on the health of doctoral mathematics departments. Despite the substantial support that is provided by federal granting agencies, far greater support for mathematics research is provided by America's colleges and universities. Foremost among them are the research universities, whose support includes the employment of both faculty with a substantial research mission and large numbers of graduate students who teach while pursuing a doctoral degree.

This approach—investing in research through America's colleges and universities—has led to enormous achievements both in education and in mathematics research. The past five decades have been a particularly successful period for American mathematics, with increasing enrollments and public support fueling striking advances in mathematics research.

But higher education in the United States is facing challenges on every front. Faculty are asked to reform teaching and to be accountable for student learning. At the same time, they are still expected to advance research frontiers and retain preeminence in the creation of knowledge. They are also asked to assume new roles in K–12 education and social programs. A fiscally conservative national climate and downsizing ethic in the 1990s has cut budgets for education, especially at the college level, along with most social programs. Universities across America face staggering financial problems, forcing them to make difficult decisions about competing priorities. At the same time, there has been an erosion of public confidence in higher education and public respect for research scientists.

Mathematics departments throughout the nation are especially feeling the strain. They are besieged by requests to reform the teaching of courses that affect almost all students in universities. Doctoral departments must nurture research programs in an increasingly competitive environment. The 1995 CBMS enrollment survey reported a substantial drop in mathematics enrollments for the first time in the survey's history. Mathematicians face a bewildering array of desires, demands, and criticisms.

Given these challenges, it is remarkable that the mathematics profession has responded to the degree that it has in the last few years. While there are many problems, there are also many successes. Building on these successes, this resource book tries to assist doctoral mathematics departments in assessing their changing environment and prospering within it. Our aim is not to criticize but to elucidate.

Task Force History

In November 1991, the Council of the American Mathematical Society charged the AMS Committee on Science Policy (CSP) to develop a science policy strategy that was consistent with the Society's mission and that addressed the issues faced by the research community. The resulting report from CSP contained these passages:

> U.S. universities and their mathematics departments share an increasing responsibility to the society in which they exist. This responsibility is met primarily by a strong commitment to quality teaching and the advancement of knowledge within the discipline, but increasingly extends to outreach activities that include the preparation of teachers, the encouragement of youth, community service, and a special obligation to encourage women and minorities to be successful in mathematics.
>
> The CSP urges the AMS to take a leadership role in the profession in advocating a rich understanding of the challenges and obligations that face our profession, especially those who teach and engage in research in our universities. While the leading model for faculty is teacher-scholar with a strong commitment to both the creation and transmission of knowledge, the AMS should promote respect for and proper rewards to those who help meet a department's total mission through focused effort in teaching, research, or outreach activities.
>
> The CSP advocates increased attention by departments to educational reform and revitalization of the mathematics curriculum, as well as to activities that encourage and nurture undergraduate students, including increasing their understanding and appreciation of mathematical research and the connections of mathematics to other disciplines and to society's needs.

Among the recommendations that the CSP made to the Society was the following:

> In order to help departments meet the broader range of responsibilities advocated by the AMS, the CSP recommends that the AMS take an active role in support of mathematics departments, with a special emphasis on supporting the needs of Ph.D. granting departments, by helping departments make the case for adequate resources from their colleges and universities. The CSP makes the following recommendations designed to support mathematics departments and the chairs who lead their departments:
>
> . . .The CSP supports the formation of a Task Force on Resource Needs for Excellence in Mathematics Instruction as proposed by the Long Range Planning Committee.

The full CSP report can be found in the November 1992 issue of the *Notices* of the AMS.

In 1992 AMS President Mike Artin appointed the ad hoc Committee on Resource Needs for Excellence in Mathematics Instruction and gave it an ambitious charge to:

- Identify the operational issues affecting doctoral-granting mathematical sciences departments.
- Conduct an analysis of the available information on these issues.
- Articulate the role of the mathematical sciences within academe and the mission of the university.
- Make recommendations on the resources needed by doctoral departments for excellence in mathematics instruction.
- Produce a cogent report for use by mathematical sciences departments and university administrations in planning and allocating resources.

Our committee got off to a slow start, in part because of the resignation of our first chair and in part because our activities were limited by a lack of resources. In late 1993 Mort Lowengrub, Dean of the College of Arts and Sciences at Indiana University, agreed to become chair of the committee. Under Lowengrub's leadership the name of the committee was changed to the AMS Task Force on Excellence in Mathematics Scholarship: Assuring Quality Undergraduate and Graduate Programs at Doctoral Granting Institutions.

The work of the Task Force increased pace in 1994 when we received partial funding from the Exxon Education Foundation, and again in 1995 when we received funding from the National Science Foundation. The support of these two Foundations is gratefully acknowledged.

Starting in August 1994 and extending through November of 1996, the Task Force held a series of 14 focus group discussions to identify the critical issues facing departments of mathematics at Ph.D.-granting institutions, as well as to gain insight into the many ways that departments are responding to the issues they face. Most of the focus group discussions (9) were held with chairs of mathematics departments at Ph.D. institutions. Other discussions were held with college deans (3), Project NExT Fellows, and department chairs at institutions that do not award the Ph.D. degree. Many readers may find the summaries of the focus group discussions the most valuable part of this book.

During the 1996–97 academic year, the Task Force also made five site visits to departments that had repeatedly been mentioned as being successful in both research and various aspects of their instructional program. The reports of those visits in Part III are not meant to hold these departments up as models, nor is there any attempt to discover any weaknesses they may have. Rather, they describe some successful practices that may suggest effective strategies for other mathematics departments.

Task Force Goals

Initially, the focus of our Task Force was helping a mathematics department make the case to get the dean to support the mathematics department better. Indeed, many chairs hoped the Task Force could produce a one-page fact sheet with information about similar mathematics departments that could help them persuade their dean to provide more support for their department.

Unfortunately, the message coming from many department chairs was that their deans were very unsympathetic to giving resources to their departments despite rising enrollments. Many chairs felt deans treated their mathematics department as a "cash cow", teaching large numbers of students at a low per-student cost. It soon became clear that convincing a dean to provide needed resources required a mutual understanding between the dean and the department of the mission of today's mathematics department and how that mission fits in with the overall mission of its university.

The goals of the Task Force expanded as we drew up a list of critical issues that departments needed to address; for example, developing strategies for implementing recommendations from recent national reports. While the work of the Task Force continued to reflect its original focus, we also saw its mission expanding.

Eventually, we recognized that our goals had become so ambitious that effectively accomplishing them all was not realistic. Our work then began to focus on a narrower set of core issues and recommendations, which were guided by an appreciation of the remarkable variety of doctoral departments. One central finding that impacted most of our Task Force's agenda was: A key to protecting and strengthening a doctoral mathematics department and its research programs is to pay proper attention to the instructional side of the department's mission. Consequently, one objective of this book is to convince research departments that they should value quality instruction not just because of its importance to the mission of the university, but also because of its importance to the overall health of a research mathematics department.

There is one potential topic that we did not address at any time: offering suggestions to departments about how to improve their research programs. Some may view this as strange for an organization that represents research mathematicians. There are reasons for this omission. First, several highly qualified panels have recently addressed this issue (see the National Research Council's "Renewing U.S. Mathematics: A Plan for the 1990's", 1991, and "Renewing the Promise: Research-Intensive Universities and the Nation", 1992, issued by the President's Council of Advisors on Science and Technology). Also, many chairs told the Task Force that deans today are unresponsive to appeals for resources that are based primarily on the need to enhance research excellence. Thus, while resources for research and doctoral training were always on the Task Force's mind, we came to believe that greater attention to high-quality instruction is the critical issue today for sustaining and enhancing high-quality research.

Why Doctoral Departments?

There are several reasons for our focus on doctorate-granting mathematics departments. First, doctoral departments produce most of our future mathematicians and much of our mathematics research. The health of these departments is important to the overall health of American mathematics. How new faculty educated in these departments view their professional responsibilities impacts all of higher education.

Second, there are a number of ways in which the instructional environment in a typical doctoral mathematics department is different from that of a department whose highest degree granted is either a bachelor's or master's degree. The most obvious ways include the heavy focus of the department on research and doctoral education. Such departments are far more likely to rely heavily on the use of graduate students as teachers (or teaching assistants). They also are more likely to rely on a large lecture format as a means of teaching large numbers of freshman and sophomores.

Third, research universities now stand accused of failing to do an adequate job (much less an outstanding job) of educating undergraduates. More broadly, large portions of society question whether universities, and especially research universities, are meeting the needs of society. The Mathematical Sciences Education Board's "Report of the Task Force on Teaching Growth and Effectiveness" argues that universities must do a better job of explaining—to themselves and to the public—exactly what it is they contribute to society. It further says that "faculty need to demonstrate the effectiveness of their educational work."

A much stronger indictment of research universities comes from the 1998 Carnegie Foundation report of the Boyer Commission, "Reinventing Undergraduate Education: A Blueprint for America's Research Universities". The report says, "Universities are guilty of an advertising practice they would condemn in the commercial world. Recruitment materials display proudly the world-famous professors, the splendid facilities and the ground-breaking research that goes on within them, but thousands of students graduate without ever seeing the world-famous professors or tasting genuine research."

It would be easy enough to reject such criticism as unfair, to state unequivocally that our Task Force believes that most doctoral mathematics departments are already doing a good job in teaching undergraduates. A premise of this book is that it is a much wiser idea to take a clear look at ourselves, taking stock of our strengths and identifying weaknesses that need to be addressed.

Increasingly, doctoral mathematics departments (and many departments in other disciplines) are challenged to defend their programs and to "do more with less". It is at least popular wisdom that departments and universities are in a time of change that is more rapid and significant than any of us have seen over the past thirty years. All too often, department leaders find themselves unprepared for many of the challenges they face. In large part, it was this quandary that led the AMS Long Range Planning Committee and the Committee on Science Policy to recommend that the AMS establish this Task Force to help departments with these problems.

All mathematics departments in this country share many concerns and aspirations, but they also have differences. The work of our Task Force was directed largely at the particular burdens and responsibilities of doctoral-granting mathematics departments.

All of our work assumes the centrality of research in the mission of doctoral mathematics departments and assumes that research and education are essential to one another. In most doctoral departments there is a clear vision for research excellence. While some departments may be struggling to achieve that vision, most have a clear understanding of a plan for doing so. There is therefore no attempt here to consider how to directly enhance the research life of a department, or to improve the research faculty, or to expand (or contract) research areas. Here too, departments differ greatly, and there are no easy answers to complicated problems.

At one time we hoped to make definitive suggestions as to how to respond to all of the critical issues facing doctoral departments. This was far too ambitious. While the information in this book can help a department determine how it wishes to respond to the variety of reform efforts that have been issued, the Task Force does not presume that there is one set of recommendations about institutional mission and instructional excellence that every mathematics department can use.

Even when focusing on doctoral departments, the differences between private and public institutions and between those at the levels the AMS refers to as Groups I, II, and III are often significant. (See Appendix B for a list of universities in these groups.) The differences between mathematics and applied mathematics departments are even more significant as, for example, in the distribution of instructional workload among graduate, upper-division, and lower-division courses. However, our concern for understanding the mission of one's institution and responding to it appropriately is important to all departments. It speaks to the differences in departments. Attention to this concern helps every department look at this book in the right way.

Chapter 2
The Environment in Which We Work

One of the most important responsibilities of a department's leadership is to position the department to receive at least its fair share of the resources available to the university. In order to meet this responsibility, it is first necessary to understand the environment in which the department and the university operate.

For background, a brief summary of changes in this environment over the past fifty years is helpful. The second half of the twentieth century has been a golden age for academic research in the United States. Many of today's academic leaders entered the professoriate when universities and colleges were fully funded and growing rapidly and doctoral programs were multiplying. Initially, mathematics departments were beneficiaries of cold war policies that put a premium on engineering and other mathematically based disciplines. Later the growing importance of quantitative reasoning throughout science and business kept mathematics enrollments growing. In the 1968 National Academy of Sciences report of the Committee on Support of Research in the Mathematical Sciences, demand for new mathematics Ph.D.'s was projected to grow to 2,000 per year in 1972 and eventually to 3,000 per year.

This emphasis on research and graduate education did not exist in the first half of the century, when most universities had an undergraduate teaching orientation and many of their mathematics faculty lacked a Ph.D. Beginning in the 1960s, university faculty became more focused on research and doctoral training. Even presidents of four-year colleges wanted their faculty to be publishing research papers, along with teaching 12 hours a week. University of California president Clark Kerr built up the UC universities beginning in the 1950s with the widely copied strategy of recruiting top research faculty with reduced teaching loads and high salaries. Before long, only research mattered in promotion and salary decisions. Today, after a sustained period of shrinking budgets during the first half of the 1990s, leaders in higher education are struggling to find an appropriate balance between research and teaching.

At a panel at the AMS 1993 annual meeting, William Kirwan, then president of the University of Maryland at College Park and now president of The Ohio State University, suggested an appropriate title for this Task Force might be "How Do Mathematics Departments Survive during a Time of Diminishing Resources and Declining Public Support?" He went on to say that universities "have needs and demands for expanded activities that far outstrip available resources," and "the only thing falling faster than our resource base is public understanding of and support for the work we do at research universities."

President Kirwan went on to say that the lack of attention given to undergraduate education is the cause of much of the criticism of research universities and quoted Derek Bok, president emeritus at Harvard, as saying the lack of attention to undergraduate education, primarily at research universities, was the number one issue causing the decline in public trust of higher education. He also quoted the following warning from Richard Atkinson, President of the University of California System, "... research universities should lead the way by restoring the balance between teaching and [research] ... the continued greatness of the American research university depends on ... an equilibrium between the three missions of its charter ... the propagation, creation and application of knowledge. When the balance goes awry, the entire edifice erodes."

More recently, in 1997, the Council for Aid to Education released a report titled "Breaking the Social Contract: The Fiscal Crisis in Higher Education", in which they wrote, "Our central finding is that the present course of higher education --- in which costs and demand are rising much faster than funding --- is unsustainable." In an open letter to faculty, the president of one public university recently identified three trends in his state:

- annual decreases in the proportion of the state budget allocated to the university,
- increases in tuition limited approximately to the rate of inflation,
- an ever increasing percentage of the operating budget absorbed by salaries.

It is encouraging to note that in the fall of 1998 many state universities reported significant increases in enrollment, and some private universities are reporting an increase in the quality of their applicants. Still, some version of the trends reported above are likely present in most states. As states struggle to fund other priorities (health care, K–12 education, prisons) and respond to various forms of taxpayer revolts, support for higher education becomes a lower priority. Private universities face a different set of problems, but they too find themselves unable to turn to traditional sources of income, such as tuition increases, to meet their need for increased revenue.

Thus, higher education, and especially research universities, face both a resource problem and a problem centered on the concerns of many that we are not meeting the needs of society. University administrators are increasingly responding to their resource problems by making hard choices as to which programs will continue to receive the support necessary to pursue excellence and which will find their support reduced significantly. University administrators are also responding to the criticism they hear by pledging greater attention to undergraduate education and the needs of the communities which support them. Mathematics departments and their leaders would be wise to pay attention to these trends.

One national initiative is trying to blend a commitment to undergraduate education with a partnership with K–12 education. Called the "P–16 Initiative", it has the strong support of The Education Trust and the National Association of System Heads. In eighteen states both the state commissioner of education and the president of the state's higher education system have formed a partnership to create a "seamless" education system from (pre)kindergarten to the bachelor's degree (i.e., grade 16). This is a standards-based initiative that places an emphasis on aligning state standards for high school graduation with college entrance requirements and also places a new emphasis on teacher preparation.

The National Science Foundation has thrown its support behind its report "Shaping the Future: New Expectations for Undergraduate Education in Science, Mathematics, Engineering, and Technology". This report, authored by Mel George, professor emeritus of mathematics at the University of Missouri-Columbus and president emeritus of St. Olaf College, sets forth one overriding goal: All students have access to supportive, excellent undergraduate education in science, mathematics, engineering, and technology, and all students learn these subjects by direct experience with the methods and processes of inquiry.

Taking Stock

Assume for a moment that you are the chair of a doctoral mathematics department and you plan to meet with your dean to discuss the department's strengths, needs, and priorities. You should examine the university's priorities and consider the work of your department. To protect the resources you have in times of budget cuts or reallocations or to seek increased resources for the department, you must match what you are accomplishing with the mission and priorities of the university. With the forums available to you, you should also take an active role in helping the institution determine its priorities. In particular, you should never tire of reminding your administration that the existence of your doctoral program and your research efforts are defining characteristics of the university.

As documented in the tables at the end of this chapter, at all but a small number of the very best doctoral programs in private universities there is a strong correlation between the size of the undergraduate instructional program and the size of the graduate program. Thus, the opportunity to build a quality research and graduate program depends in part on the size of the undergraduate program. It is not too great a stretch of the imagination to believe that inadequate attention to undergraduate education will place research and graduate education at risk. This is, of course, consistent with the arguments of Kirwan and others that we must place greater emphasis on undergraduate education if we are to protect America's commitment to the research university.

More than anyone else, it is your responsibility to convince your administration that an excellent undergraduate mathematics program is worth paying for. Indeed, you must remind them that quality undergraduate mathematics instruction, with or without innovations, is a labor-intensive activity.

One of the strongest arguments available to mathematics departments is a combination of the centrality of the discipline together with the size of the instructional program. Many departments can point to the fact that they provide as much as 7 percent of all instruction at the university. Virtually every undergraduate takes at least one course from the department, and in many undergraduate colleges, in any given semester, between 25 percent and 45 percent of all students are taking a mathematics class. If retention of undergraduate students is important and if the university is trying to make a greater commitment to undergraduate students, then the mathematics department is central to their success or failure.

At the same time, this argument can backfire inasmuch as 88 percent of all instruction offered by the department is at the freshman and sophomore level. Deans who are looking to trim costs often ask whether this is instruction that can be offered by lecturers and other part-time instructors. An analysis of many departments reveals that only about 4 percent of the students taking mathematics courses are at the graduate level and another 4 percent are students who are majors in the department. Evidence of faculty interest and involvement in calculus instruction and in the offering of high-quality courses, both as part of the university's general education efforts and to meet the mathematics instruction needs of future teachers, engineers, and scientists, is important to preserving the link between the size of the instructional program and the size of the research faculty.

The essay, "A View from Above", written by Professor Jim Infante, dean of arts and sciences at Vanderbilt University, and included in this report, provides valuable insights into how your proposals to your dean will be judged. In planning for your meeting with your dean, it might be appropriate to take stock of your department's strengths and weaknesses with respect to the following aspects of the department's work:

- Research
- Interdisciplinary work
- External funding
- Graduate education
- Remedial and other precalculus instruction
- Calculus
- General education
- Teacher preparation
- Majors
- Outreach
- Diversity

Research. As noted in the previous chapter, this report consciously says little about research issues. However, in the context of talking to the dean, the following comments on research seem appropriate. Mathematics department chairs often find it difficult to defend their research program, in part because of comparisons to other science disciplines. Physicists, biologists, engineers, etc., tend to publish many more papers than mathematicians and attract much larger external funding. They can fund more graduate students and postdocs than almost all mathematics departments. Mathematics departments also suffer from inadequate media coverage of their research and an absence of ways to document the quality of their faculty based on various forms of public recognition. For example, mathematics lacks anything comparable to the "fellows" designation of the American Statistical Association. Nonetheless, mathematical research is widely respected for its deep intellectual nature. Most educated people are aware that the world we know today could not exist without the tools and knowledge that grew out of mathematical research of the past. While other disciplines sometimes chide mathematicians for being too far removed from real-world problems, they still expect that some of this far-removed thinking will prove invaluable in the future.

There are a number of National Research Council reports and articles in the AMS *Notices* that document important practical uses of contemporary mathematics, such as the importance of group and number theory in cryptanalysis, differential geometry in unified field theory, wavelets in image compression, scattering theory in magnetic resonance imaging. Thus, the importance of curiosity-driven research lies both in the elegant and powerful mathematical theories it creates and in "the unreasonable effectiveness" (to use Eugene Wigner's phrase) of this mathematics in science and engineering. The leadership in a mathematics department needs to educate the university administration and colleagues in other departments about the mathematical research enterprise—its vast spectrum, its interconnections, and its impact.

Interdisciplinary Work. Research and education today are assuming an interdisciplinary character. Unfortunately, as mentioned in the focus group discussions with deans (Chapter 6), all too often deans see their mathematics departments as "insular". However, most mathematics departments today have a substantial amount of interdisciplinary collaboration going on in research and/or in teaching. Sometimes it involves, say, some physicists at another institution who work with some mathematicians who in turn collaborate with faculty in your department. It is important for a department chair to be fully informed about all these activities. Also, as noted in the previous paragraph, more and more of the problems mathematicians work on today have important connections to other disciplines. There is much to be gained by trying to build additional bridges to faculty in other departments on campus (this can be done both through research partnerships and through education initiatives) and putting to rest the charge that mathematics departments tend to be insular.

External Funding. Many university departments, especially in the sciences, are judged in large part by their ability to generate external funding. While administrators know that the external funding for mathematics research is far below that in the sciences, total funding and the percent of mathematics faculty with external funding are still important to administrators and to mathematics departments. Moreover, substantial external funding does exist in applied mathematics and mathematics education. Deans are much more likely to commit university funds to research or educational initiatives when there is evidence that their support will help secure funds from some outside source.

Beyond funds for faculty summer salaries, computers, and graduate assistant stipends, external support can play a vital role in the quality of departmental life.

Most universities return to departments a portion of indirect cost funds, which can be an important, though modest, source of discretionary funds. Because of the importance of external funding, it is valuable for departments to create a culture of proposal writing. This includes an attitude, common in other disciplines, of not treating a rejected proposal as a failure, but rather as a challenge to learn from the referee reports, revise, and resubmit the proposal.

It will be increasingly important for departments to actively seek donations from alumni, businesses, and foundations to provide for scholarships, educational initiatives, and research support. Even modest discretionary funds to improve such "quality-of-life" issues as visitor support, social activities for students and faculty, and travel can be a bracing tonic to a department. Close coordination with your university's development office can be crucial to your success.

Graduate Education. While graduate education is a part of their responsibilities that research mathematicians care about deeply, it is seldom a basis for an argument for more resources. In most universities, the number of Ph.D.'s awarded in mathematics is small compared with the number in education, business, psychology, and numerous other disciplines. Thus, an argument based on the size of the program pales by comparison to many others. If, however, the department can offer evidence that the department's graduate program is of particularly high-quality for the university, then the graduate program becomes a department strength the university is pleased to support.

Departments who choose to develop interdisciplinary programs or professional master's programs to meet specific needs (financial mathematics, industrial mathematics) are broadening their mission and advancing their university's ability to provide graduate training in emerging professional specialties. In turn, such programs can help strengthen their university's commitment to the graduate program in mathematics.

Administrators, especially at public institutions, are impressed if departments can publicize the diverse, good jobs their M.S. and Ph.D. students obtain. Departments should provide their graduating Ph.D.'s with training in the nuts and bolts of job hunting, including the preparation of applications, how to target different types of institutions, and trial interviews. Chairs should keep records as to where their graduates take jobs and should assess whether their jobs match the education they received. In addition to knowing which colleges and universities hired their recent Ph.D.'s, it is important to recognize that increasingly students at the graduate level are taking jobs outside academia. For example, a large number of quantitatively based careers in emerging new fields such as quantitative finance require a traditional education in a discipline such as mathematics. If your department is successfully placing students in business and industry, such information may be welcomed by administrators seeking to defend arts and sciences budgets before state legislators.

In considering ways to improve the graduate education offered by your department, it is important that faculty do not change the best part of the graduate experience, the Ph.D. thesis. Project NExT fellows were very positive about their training to do mathematics research. Industrial employers also praise the value of an in-depth experience of working on hard problems. They praise the "analytic thinking skills" that graduates develop, and advise against making significant changes in this core doctoral experience.

Remedial and Other Precalculus Instruction. As many chairs told the Task Force, remedial instruction and, more generally, precalculus instruction pose a significant problem for many departments. In places where there is a large demand for remedial instruction, it can drain resources from the department and time from the department administration.

If remedial instruction results in numerous complaints to the dean or provost, it is surely a matter that must be dealt with before any department priority receives a warm reception from the dean. On the other hand, it is unlikely that the department's needs in this area will result in significant new resources because of the perception that remedial instruction is not very important to the university and that such courses can be taught cheaply. If, however, the department can link precalculus instruction (including remedial courses) to the university's retention efforts, then success in this area can open doors to discussing other priorities.

Calculus. Calculus instruction, on the other hand, is central to many disciplines on campus and is often viewed as the key to whether students will be successful as engineers, scientists, etc. If your administration is convinced that your department cares deeply about calculus instruction and is striving to provide high-quality calculus instruction, they will almost certainly work to find the resources you need for this purpose. In particular, a number of departments have found their university administration supportive of curriculum projects designed to improve calculus instruction at their university.

General Education. In recent years many universities have launched a "general education initiative" seeking some common core of knowledge for all students at their university. Many mathematics departments responded by putting energy into the development of new "liberal education" or "general education" courses for majors in the arts, humanities, and social sciences. Basically, this amounts to developing meaningful (but accessible) mathematics courses for students who will not take some form of calculus from the department. Whenever this fits a university priority, it becomes a basis for arguing for more resources if the department is responding creatively to the university's initiative.

Teacher Preparation. Teacher preparation is an area of collegiate instruction that traditionally has been a low priority in research universities. The recent attention paid to the success (or lack of success) of K–12 students in international comparisons such as the Third International Mathematics and Science Study (TIMSS) has led to a significant increase in attention to teacher preparation and to the continuing education of current teachers. NSF has attempted to focus greater attention on teacher education with its "Shaping the Future" report. The P–16 Initiative mentioned earlier is causing teacher education to be a priority for university presidents who would be hard pressed to show how it was a priority five to ten years ago. Mathematics, of course, is right in the middle of any national priority in K–12 education and the preparation of teachers. Departments that become seriously involved in this aspect of their mission are likely to see benefits for all aspects of department life, and departments that ignore teacher education may suffer.

Mathematics Majors. The education of mathematics majors is, next to graduate education, the part of our instructional mission that appeals most to mathematicians. But our achievements in this aspect of our work are not likely to be perceived as particularly important to university administrators. On most campuses, the number of mathematics majors is tiny by comparison with those in biology, psychology, engineering, business, or education. In a review of doctoral mathematics departments, only five are regularly graduating over 100 majors a year and only sixteen are graduating over 60 majors a year.

Despite the relatively low numbers of students majoring in mathematics, a department's success with majors might be the basis for increased resources if the department can argue that its majors tend to have higher academic credentials than the typical student on campus or that graduates are successful in obtaining outstanding jobs. This argument will be particularly effective on a campus that places a high priority on recruiting outstanding students.

Outreach. Both private and public universities recognize a need to be good corporate citizens in their state or community. Some universities, especially land grant universities, identify outreach (or service) as a significant part of their mission. Mathematics departments have an excellent opportunity to contribute to this part of the university's mission by becoming involved in professional development programs for teachers or by developing special programs for students in the K–12 school system. Such programs can bring very positive attention to your department and college and can often be the basis for proposals for external funding. Responding to the nationwide interest in distance education may be another way for a department to become involved in an outreach activity.

Diversity. Most universities have identified campus goals that may be broadly identified as promoting diversity in our society. These goals can focus on either the success of students or on faculty hiring. Mathematics departments can make a big contribution to their university by developing programs that improve the success of underrepresented minorities in college or that increase the number of women and minorities who are successful in mathematics-based disciplines or in graduate school. Some excellent examples are discussed in Part III of this book.

We Are Not All Alike

As indicated in the first chapter, the Task Force focused its work and its recommendations on doctoral mathematics departments, the ones commonly referred to as Group I, II, and III departments. As noted earlier, they share a common mission, common problems, and common approaches to much of their work. Even the so-called Group V, applied mathematics departments, are very different from the Group I, II, and III departments. For example, these applied mathematics departments have an instructional mix very different from the profile of doctoral departments presented below.

The following table gives the instructional profile for private and public doctoral mathematics departments based on the total student enrollment at each level. Thus, a graduate student who takes three mathematics classes would be counted three times. The Group I departments have been split roughly in half:

GrIA contains the highest-rated departments, and GrIB consists of the remaining Group I departments. While there are 177 Group I, II and III departments, this table uses data from only the 112 departments that submitted data to the AMS-IMS-MAA Annual Report for both fall 1992 and fall 1997. The data below is for fall 1997.

Percent of Total Student Enrollment at Various Levels								
	PRIVATE				**PUBLIC**			
	GrIA	GrIB	GrII	GrIII	GrIA	GrIB	GrII	GrIII
Remedial	0.0	0.7	1.5	1.1	8.9	8.1	7.3	16.9
Precalculus	2.4	5.7	4.0	14.3	11.4	21.7	23.6	19.6
1st Year Calculus	47.2	46.0	45.3	38.8	39.9	34.5	30.6	23.5
Stat or Comp Sci	3.4	9.1	15.9	9.2	2.9	2.9	4.0	8.3
Courses for Majors	16.4	18.7	21.3	21.2	18.9	18.8	13.7	10.4
Other Und Courses	20.4	15.6	9.0	12.2	13.8	11.0	17.3	17.9
Graduate Courses	10.3	4.2	3.7	3.2	4.2	3.1	3.4	3.4

Among the items that stand out are the following:

- Only at the highest-rated 10–12 private institutions (i.e., Group IA) is graduate student enrollment a significant percent of total enrollment.
- Private institutions offer virtually no remedial mathematics and very little precalculus instruction.
- Even the highest-rated public universities have a significant remedial mathematics instruction problem, although it is much greater at Group III institutions.
- First-year calculus is a very large part of the workload at private universities. At public universities, calculus is a much larger share of total instruction for GrIA departments than it is at Group III institutions.

Very few doctoral mathematics departments continue to have instructional responsibilities in the area of computer science. Statistics is more likely to be important for mathematics departments at private universities and at smaller Group III public university departments.

Average Number of Course Registrations – Fall 1997								
	PRIVATE				**PUBLIC**			
	GrIA	GrIB	GrII	GrIII	GrIA	GrIB	GrII	GrIII
Undergrad	1,717	1,841	1,738	1,646	7,789	5,938	4,696	3,437
Graduate	197	81	66	54	338	192	165	121
Total	1,914	1,922	1,804	1,700	8,127	6,130	4,861	3,558

For private universities there is little variation in the sizes of the undergraduate programs for different categories of departments. With the exception of the GrIA departments, the size of the graduate program seems to be driven by the size of the undergraduate program. The GrIA private departments appear to have a significantly larger graduate program that cannot be explained by the size of the university.

For public universities there is a remarkable relationship between the size of the undergraduate program and the departments which had the highest faculty rating in the most recent NRC rankings. Here, too, the size of the graduate program is clearly related to the size of the undergraduate program, with the exception of the GrIA departments.

What conclusions can be drawn from this information? Certainly, to some degree, the size of the faculty and the size of the graduate program are related to the size of the undergraduate program. Only at the highest-rated mathematics departments are graduate enrollments more than 4 percent of total enrollments, and even in those departments the percentages are not large. The message for all of us is clear: A commitment to high-quality undergraduate education is not only the right thing to do but is necessary if we are to protect research and graduate education in our research universities.

Chapter 3
What We Learned

We learned much throughout the course of our work. The extensive comments of chairs and deans in the focus group discussions showed both the nature of the problems we face and the difficulty of achieving simple solutions. Our indepth site visits (as well as shorter visits to other departments) showed the ways in which some of these problems are being addressed in specific situations. In our meetings we considered all these comments and observations, and we tried to draw conclusions. This chapter describes those conclusions.

1. The nature of academic life is changing.

Most universities find themselves in a period of retrenchment. Reallocation or budget cuts are far more common for universities than the periods of rapid growth that many universities experienced in the '60s, '70s, or '80s. According to the Council for Aid to Education report "Breaking the Social Contract: The Fiscal Crisis in Higher Education", "the present course of higher education—in which costs and demand are rising much faster than funding—is unsustainable." Indeed, this report goes on to say, "What we found was a time bomb ticking under the nation's social and economic foundations: At a time when the level of education needed for productive employment is increasing, the opportunity to go to college will be denied to millions of Americans unless sweeping changes are made to control costs, halt sharp increases in tuition, and increase other sources of revenue."

One need only look at the sweeping changes in health care or at the many states suffering through various kinds of citizen tax revolts to realize that significant change in higher education is possible and that one ignores this possibility at great peril.

Within academia there is growing criticism of research universities for neglecting undergraduate education. The report of the Boyer Commission on Reinventing Undergraduate Education sets out an "Academic Bill of Rights" in an effort to describe the undergraduate education that all students should be guaranteed at a research university. University of California president Richard Atkinson (quoted in Chapter 2) has called for restoring the balance between teaching and research. The NSF report "Shaping the Future" is concerned that "All students have access to supportive, excellent undergraduate education in science, mathematics, engineering, and technology."

The mathematics department that does not help its institution to accommodate changes in higher education may find some of its resources reallocated to other sectors of the institution.

2. Departments must invest effort into understanding their university's mission and priorities and then positioning themselves to meet those priorities.

If a university is concerned with retention, attracting honors students, raising academic standards, improving the success rate of minority students, or providing a common core of learning as part of a general education initiative, then surely the department should be asking itself whether it is contributing appropriately to these efforts. If the institution is interested in teacher education or creating a "seamless educational system, from pre-kindergarten to grade 16," then a department should be asking what role it should be playing in this effort.

We do not suggest that a department must blindly follow institutional directives. Faculty can take an active part in helping their university set priorities by balancing constructive criticism with support of institutional goals. For example, rather than merely protesting increased attention to undergraduate or K–12 education, faculty can engage in debate about the most effective means to achieve these goals while maintaining other institutional needs. It is neither appropriate nor effective to block change while offering no constructive alternatives.

When arguing against any ill-considered changes, faculty still need to try to find common ground with their administration and align their department's priorities as much as possible with their institution's priorities. Only by making a meaningful contribution to their institution's priorities is a department likely to receive additional resources.

3. A strong commitment to high-quality undergraduate instruction and to other educational activities should be an integral part of the mission of every doctoral mathematics department.

Undergraduate education is becoming more important in defining the mission of research universities. Thus, many mathematics departments will need to invest more resources (people, time, and money) and intellectual creativity in undergraduate education. Inadequate concern for a department's undergraduate instructional program is sure to bring increased criticism. On the other hand, a department that earns a reputation for excellence in teaching undergraduates generally finds that this pays clear benefits in terms of the resources that are allocated to the department.

The goal of our Task Force is to strengthen research mathematics departments. Indeed, we have spent most of our academic lives in research departments. In the course of serving on the Task Force, we learned of many outstanding examples where departments are successfully responding to the challenge of enhancing their undergraduate program while remaining committed to the development of excellence in research and graduate education.

At the same time, we are concerned that many deans believe their departments have marginalized undergraduate instruction, especially the 60 percent of their instruction at the level of calculus and below. It is in our own best interest to reexamine our commitment to this part of academic work and to be sure that it is an integral part of what we value.

4. Strong leadership is essential to department success.

It is surprising that this statement is not a statement of the obvious, but it is not. Ideally, a department will have a strong department chair who is backed by a solid leadership team (e.g., vice chair, graduate chair, undergraduate advisor) and who has the backing of the senior faculty within the department. Most important of all is a strong department chair who is an effective advocate for the department.

Many mathematics departments appear to fear strong leadership rather than value it. This viewpoint is presumably an outgrowth of a belief that the department is the master of its own destiny; that the department's greatest concerns are internal; and that faculty must guard against the possibility that a strong leader, especially a strong department chair, will impose his or her biases on the department.

Our Task Force believes that much greater challenges and opportunities come from outside the department. In a climate of change, it is important for the department to have a strong chair who can articulate the views of the faculty to the dean (or other administrators) and who can work effectively to secure needed resources for the department.

Repeatedly, our Task Force saw a correlation between a strong department chair and a successful department. Whether the university environment was one of competing for new resources, reallocating constant resources, or determining where budget cuts should occur, strong leadership mattered. At the same time, as our Task Force spoke with chairs it learned of far too many occasions where department culture did not assign appropriate value to the job of department chair or worked to limit the effectiveness of a chair.

Department chairs repeatedly told our Task Force that it took a couple of years to fully understand how their university worked and how decisions were made. The tendency of departments to prefer "rotating chairs" often resulted in chairs leaving their positions just as they were finding themselves able to speak effectively for their departments.

In focus group discussions, a number of deans told our Task Force that they perceive an excess of internal strife in mathematics departments: between pure and applied mathematicians, and between traditional and reform approaches to instructional philosophies at the undergraduate and K–12 levels. Effective department leadership—involving more than just the chair—can create and maintain a healthy environment for the discussion of differing viewpoints, an environment of mutual respect that maximizes both sides' common concerns for quality research and education.

5. Successful departments have established credibility with the university administration.

In site visits and discussions with chairs there was a common theme: successful departments had earned the confidence of their university administration. They understood that the centrality of mathematics carried with it a responsibility to meet the needs of the campus. These chairs were able to cite examples of how they had taken the lead in responding to the challenges faced by a mathematics department in a research university. They were meeting their responsibility for the mathematics education of students from all disciplines and of widely varying abilities. They were actively involved in leadership positions around the campus, and faculty research was making notable contributions both within the discipline and across disciplinary boundaries. Successful departments set goals, strategies, and plans for contributing to the overall mission of the university.

Department leadership is important in establishing the credibility of the department. Deans and provosts must understand the department's priorities, and they must trust the department chair to provide accurate information about the department and to communicate their concerns to the department.

Unfortunately, administrators often assume that responding to complaints about mathematics instruction is a necessary aspect of their position. By demonstrating its commitment to and competence in providing high-quality undergraduate teaching, a mathematics department will gain important leverage in seeking support for other department priorities.

6. The need to defend research will increase.

In very strong departments, say the highest-rated twenty-five research departments in the NRC rankings, institutional commitment to research may be secure. Most other departments may increasingly find themselves needing to defend mathematics research. This can be difficult. Seldom are mathematicians prolific publishers in comparison with their science colleagues. The size of research grants pales by comparison with those in lab sciences and engineering. Mathematics research does not have media attention and public understanding on a par with research in biochemistry, physics, agriculture, etc.

As discussed in the previous chapter, there are new calls for accountability and "measures of productivity", and the need to explain and defend mathematics research will increase. As universities look for things that can be cut or reduced, activities that university administrations do not understand become prime candidates for what they will stop doing.

As the David Report demonstrated, the mathematics community, in concert with friends in other disciplines, can make a strong case for the importance of fundamental research in mathematics and its centrality to many advances in science. Mathematicians need to become more conscious of the need to promote the value of mathematics research to faculty outside mathematics, to administrators, and to the general public.

7. Depending upon the mission of the department and the university, a significant educational outreach program may be appropriate.

Increasingly, universities and departments are challenged to make broader commitments to serve the community in which they are located. Mathematics departments and mathematics faculty can make a major contribution by becoming involved in teacher preparation or continuing education for teachers, enrichment programs for K–12 students, or efforts to help minorities succeed in mathematics. Again, the Task Force found examples of departments that are active in outreach, also have a strong commitment to undergraduate instruction, and continue to excel in research and doctoral training.

8. Issues of diversity are increasingly important to universities and to the profession.

American colleges and universities play a key role in maintaining a classless American society by providing opportunities for citizens to advance economically, professionally, and socially, consistent with their ability and commitment to hard work. Thus, American colleges and universities have always had a special responsibility to society.

One of the most challenging issues faced by higher education is the need to provide meaningful educational opportunities for minorities, especially African Americans, Hispanics, and Native Americans. In science, mathematics, engineering, and technology, we are faced with the added responsibility of providing increased opportunities for women. It is no longer acceptable (if it ever was) for a department to adopt a passive approach of willingly teaching those who come to them but making no special effort to create opportunities and offer encouragement to underrepresented groups of students.

Most universities have identified diversity as a major campus priority. This can mean many things, but it almost certainly includes the goal of increasing the number of women and minorities on the faculty and the number of women (in science disciplines) and minorities who successfully graduate from its graduate or undergraduate programs. For some universities it also means the need to do more in terms of closing the gap between majority and minority populations in public schools.

Mathematicians argue that their discipline has a special role to play in universities because of the centrality of the discipline. This is especially true in terms of enhancing the success of students drawn from populations that historically have not been successful in mathematics and science. If departments make major contributions, they should be able to expect tangible rewards in return.

9. Most departments need to rethink the goals of their graduate program.

Graduate education is connected to the Task Force's findings about the changing environment and the increased importance of undergraduate education. According to the 1997 Annual Survey (second report), less than 20 percent of new Ph.D.'s obtained jobs at a Ph.D.-granting institution in the U.S., including jobs in statistics and applied mathematics departments. At Group I institutions, only about 30 percent of their graduates were hired by Ph.D.-granting institutions. A substantial percentage of these new Ph.D.'s were hired in postdoctoral positions or other temporary positions. It is reasonable to assume that even fewer will eventually obtain tenured positions at Ph.D.-granting institutions. Certainly this prompts the question, For what positions are we preparing graduate students?

There is some good news on this front. Project NExT, sponsored by the Mathematical Association of America, has worked with over three hundred new Ph.D.'s to help introduce them to the many aspects of their new professional life, and the Project NExT Fellows appear quite active at meetings and in professional organizations as a result. A number of departments (e.g., the University of Washington) have become involved in the Preparing Future Faculty initiative sponsored by the Pew Charitable Trust. Working across many disciplines, these programs work with graduate students to help them develop expertise in teaching as well as in research and learn about professional life at a wide variety of institutions, including two-year colleges, liberal arts institutions, and comprehensive universities. The MAA publication *You're the Professor, What Next?* offers a wealth of essays about programs designed to help prepare graduate students for the profession.

A number of departments have worked to prepare graduates who are attractive to American business and industry. It is interesting to note that in the Annual Survey of new Ph.D.'s, the number of jobs in business and industry reported was 248 in 1997. This compares quite favorably with a total that averaged about 100 in the 1980s and was only 114 as late as 1994.

The National Science Foundation's new mathematical sciences initiative, Vertically Integrated Grants in Research and Education (VIGRE), ties several of these themes together. At the graduate level its program announcement calls for restructuring of graduate education to integrate training in research and teaching, along with outreach experiences either in industry or in local schools.

Since the mid-1990s the number of first-year graduate students and the total number of mathematics graduate students at Ph.D.-granting institutions has been dropping dramatically. While this may be a "market correction" in the number of new Ph.D.'s attributable to factors beyond departments' control, it is worthwhile nonetheless for a greater number of mathematics departments to reform their graduate programs with an eye toward preparing graduate students for teaching positions in non-Ph.D.-granting institutions and in business and industry.

10. Both teacher preparation and K–12 outreach merit a greater share of the time and attention of mathematics departments.

Most research mathematicians work at institutions that produce significant numbers of teachers at both the elementary and secondary level. All too often, teacher education is in a separate school of education and largely distinct from the work of the mathematics department, and few, if any, of the tenure-track faculty are involved in teacher preparation. Many research mathematicians view courses for elementary school teachers with the same low opinion they have for courses such as precalculus. If K–12 mathematics education in the U.S. deserves criticism (and it surely has received a lot of criticism in the wake of the TIMSS reports), then a share of the blame falls to those university mathematicians who should be playing an important role in the preparation of teachers but are not. It is easy to make the case that among the most important students mathematicians teach are future school teachers—students who will each pass on the mathematics they have learned to hundreds of other young people.

Beyond the preparation of the next generation of teachers, it is likely that colleges and universities will be called upon to play a larger role in the important business of improving mathematics education in the U.S. This will require more mathematicians taking a role in the continuing education of teachers and making a contribution to the public discussion of what is taught and how it is taught. For most departments this is a fertile area for making a contribution to the university's mission.

11. Adapting to changing priorities is a continuing obligation.

Most mathematicians were educated in an environment where the job of faculty at a research university was restricted to research and teaching. Only small amounts of service were necessary to keep the department operating. The current environment requires a continual commitment to justifying the department's activities, arguing for resources, and establishing plans for the future. Today curriculum renewal, K–12 outreach, teacher preparation, and other educational activities all demand significant amounts of department attention.

As much as faculty might like to "fix the problem" and get back to life as it used to be, that is unlikely to happen. At least for the short term, this richer, more complex mission will be the order of the day in mathematics departments.

12. Department reward systems must reinforce department priorities and recognize contributions in all aspects of a department's mission.

It is a simple observation that departments must decide what professional work is important to the department's mission and then find faculty who will accomplish that work. They will succeed only if the department rewards the work it values. A preferred model for a faculty member is the teacher-scholar mentioned in the science strategy developed by the AMS Committee on Science Policy. Our Task Force endorses the CSP's call for respect for and proper rewards to those who help meet a department's total mission through focused effort in teaching, research, or outreach activities. It is inevitable that different faculty will develop differing strengths and different areas where they can make their most valuable contributions.

One mathematics department, in the top dozen in the NRC rankings, provides a striking example of revised priorities matched with a revised reward system. While devoting extensive faculty resources to innovative calculus instruction in small classes, the department's senior faculty voted to grant a named professorship to the leader of the calculus initiative and gave tenure to the head of its mathematics learning center. The department's commitment to undergraduate education and its documented impact on retention rates resulted in substantial new faculty resources for the department to expand the number of innovative calculus classes.

13. Data is becoming much more important.

Department chairs repeatedly told the Task Force that as the demand for accountability or for measures of productivity increase, they need more data. Each department needs information that helps compare the achievements of the department with similar departments across the country. While the AMS-MAA-SIAM annual reports provide much valuable information, department chairs seemed to be indicating that they need even more information to speak for their department or to know when they should be able to accomplish more with current resources.

14. While there are many problems for mathematics departments (and universities), there are also many successes.

Research mathematicians should mix honest criticism with pride in their accomplishments. Mathematics has much to be proud of, both as a profession and in the way it has addressed a number of the issues discussed here. Mathematicians have been as heavily involved in curriculum changes as any discipline in higher education. Scholars from around the world continue to come to the United States for their training in mathematics. American mathematics research continues to lead the world in many areas. Many, many students at all levels trained in doctoral mathematics departments go on to productive careers, not just in mathematics but in other disciplines as well. Mathematicians should keep these accomplishments in mind when considering changes so that they do not abandon those things that make their programs strong.

Much public criticism of mathematics education by mathematicians is aimed at enlightening the mathematics community in order to address outstanding problems. Mathematicians advocating change must take care not to criticize colleagues in public too vigorously or try to pressure them to be penitent. Criticism without balance makes it more difficult to find solutions.

The views discussed in this chapter led our Task Force to the recommendations that are presented in the next chapter. "Balance" is a critical word in all these discussions. Balance between research and teaching. Balance between sometimes conflicting institutional and departmental priorities. Balance between tradition and change; between established practices (many still valuable but some worth rethinking) and new approaches (many well intended but some unrealistic). Striking the proper balance on these issues is the biggest challenge facing the leadership of every mathematics department.

Chapter 4
Our Advice

The two previous chapters describe the environment in which doctoral mathematics departments are likely to exist over the next decade and list observations that our Task Force believes should guide the decision-making process within doctoral departments. This chapter presents our recommendations to the mathematics community, especially to the chairs and faculty in doctoral departments.

First, we offer three guiding principles that are crucial to the success of a mathematics department. They follow from the findings in the previous chapter.

- Understand the mission of the university and the role of the mathematics department in achieving that mission.
- Create an environment that encourages, enhances, and enables the creative work of the faculty and students who together make up the department.
- Obtain the resources, both human and financial, needed to accomplish the goals of the department.

As we stressed in the previous chapter, making the need to understand the mission of the university a guiding principle should not be misconstrued as suggesting that individual faculty or departments should blindly follow wherever university administrators lead. Instead, mathematics faculty, especially the department leadership, should work to become integrally involved in determining university priorities and in arguing for an institutional value system that places high priority on the core values and activities of a research university. However, to have access to these decision-making councils and to have influence in them, a department's leadership will need to have earned the respect of the university administration through its contributions to advancing other aspects of the university's mission.

If your university places great emphasis on the retention of undergraduate students and perceives the mathematics department as the greatest impediment to improved retention, then the department is unlikely to get new resources (for example, for its graduate program) until it convinces the administration that it will contribute to the retention effort. If the department's size and resource base is due in part to the need to provide precalculus instruction to large numbers of students, the department must convince the university that it accomplishes this part of its mission successfully in order to gain administrative support for the department's highest priorities.

Likewise, if the university has assigned a high priority to gaining membership in the AAU (American Association of Universities) or to improving the NRC rankings of its top departments, then strategies for strengthening the research capacity of the department will be supported by the administration.

As we indicated in Chapter 3, most successful departments have established credibility with their university administration and particularly with their dean and provost. They have done this by recognizing clearly their special position (the centrality of mathematics) and the responsibility that goes with it. They have taken the initiative to address the enormous range of challenges they face. The successful department has earned a role as a campus leader by setting and achieving goals that advance the mission of its university.

Before moving to a specific list of recommendations, we offer the following goal for consideration by most doctoral mathematics departments.

> The Department of Mathematics will be a model department whose mission includes a commitment to excellence in both research and educational activities.

There are some important caveats to offer at this point. Each mathematics department must make its own decision as to the proper balance between the department's commitment to research, to graduate education, to undergraduate education, and to other educational activities. There is no one correct model. Instead, we offer some examples where departments have made important contributions to their university through their educational activities and where it appears to our Task Force that all aspects of the department's mission have benefited as a result.

A loud message from the focus groups with deans was the perception that many mathematics departments were not giving adequate attention to their instructional responsibilities. Our goal seeks to redirect this criticism, turning a dean's concern about good mathematics instruction to a department's advantage.

Having advocated instructional excellence, we remind our readers that this book is targeted primarily at faculty who work in doctoral mathematics departments. It is already a part of the basic mission of your university and your department to have a commitment to mathematics research and to graduate education. Almost certainly your institution is, or wants to be, a Research I or Research II institution in the Carnegie Classification. Your continuing concern about this part of your mission is central to defining who you are. It is a concern for the resource base for research that motivates in part the above goal.

Recommendations for Departments of Mathematics

The following recommendations present important components for achieving the goal of becoming a model department at your university. These recommendations are interrelated. A department will have the greatest success if it considers the recommendations as a group and implements as many of them as possible.

1. Develop a plan.

- Assess your department's commitment to research, graduate education, undergraduate education, outreach, and related educational activities.
- Determine whether the balance is appropriate for your university or whether changes are necessary.
- Develop a mission statement and strategic plan that will strengthen the department and enhance its standing with administrators responsible for resource allocation.

This plan must simultaneously be faithful to the values of our discipline and responsive to the needs of your institution. It is wise to consult with your dean early in your planning process. The plan should be developed by the department as a whole and should have the broad support of the faculty. The plan should be summarized in a mission statement that is as explicit as possible. This statement will be a public document that will serve as a reference point in discussions with administrators about the utilization of current and future resources. The mission statement should maximize the strengths of mathematics and minimize any weaknesses. As noted previously, mathematics plays a central role intellectually in the educational mission of a university. It also is central in practical ways: for example, student success rates in mathematics have a significant impact on retention. While research in mathematics may not fare well in terms of external funding when compared to the sciences, administrators still recognize that there is substantial academic cachet in being able to count mathematics among their highly ranked departments.

Many universities require departments to conduct a department self-study on a regular basis (e.g., every five years). Part V of this book offers a guide that can be used for a self-study or an external review. Certainly, if a department is required to make a major investment of time and energy in an external review, it is reasonable to try to make certain that the review serves the needs of the department. Frequently an external review is carefully controlled by an administrator, and the department's greatest concern is avoiding harmful results from the review. Despite the risks, this type of review has the best chance of obtaining needed resources. Such a review is most likely to be of benefit if the department (and the department chair) are able to work cooperatively to plan the review.

Sometimes a department can get more out of a self-study that is completely controlled by the department because such a review permits the department the freedom to be honest with itself about its strengths and weaknesses. The benefit of this type of review is limited to those issues the department can affect through its own actions and resources, as upper-level administrators are likely to ignore any review in which they had no involvement.

This book and the other resources to which it refers can be useful to faculty as they assess their current department and develop a plan for strengthening the department. We hope that many departments will make strengthening their commitment to educational issues a major part of their plan.

2. Make a commitment to quality undergraduate instruction.

No single issue is more important than undergraduate instruction in determining whether research universities, especially public universities, will receive strong support from alumni, legislatures, business leaders, and the general public. We can debate endlessly whether the criticism that higher education has been getting is fair, but the fact remains that universities do not have the public support that they once had and that they certainly need.

Mathematics departments often offer as much as 7 percent of all instruction at a university and a much higher percentage of freshman- and sophomore-level instruction. Ideally, the mathematics department should be a source of pride for the quality of instruction offered by the university. Because most students find mathematics courses to be among the most difficult they must take, it takes special effort for the department to establish a reputation for excellence in instruction, but it can and should be done.

As more states struggle with mounting evidence that many students graduate from high school unprepared for work or college, greater attention is being paid to the need to invest in high-quality teacher preparation programs. Because mathematics is a large part of a K–12 education, we in mathematics departments must be prepared to do more to help prepare high-quality teachers. Some universities prepare hundreds of teachers each year, while others have no program specifically designed to prepare teachers. Clearly, the proper role for the department in this issue depends upon the university's commitment to teacher education.

3. Support outreach. Determine the department's potential role in helping its state and local community, and develop an appropriate outreach mission for the department.

Increasingly, universities realize that they cannot expect continuing support from state or local communities without making a contribution to their well-being. For a mathematics department the most obvious roles are associated with the continuing education of teachers of mathematics and outreach programs for students still in the K–12 educational system. Because of the current attention focused on K–12 mathematics education, a department that has a significant outreach program working to improve K–12 mathematics education is sure to be a source of pride for the university.

If the mathematics department has an applied mathematics group, then research collaborations with regional industry should also be possible. As one other possibility, several mathematics departments have started actuarial tracks to serve the insurance industry. Another idea is to support the university's interest in distance education by offering mathematics courses via the World Wide Web.

4. Broaden the preparation of graduate students. Prepare graduate students for their profession and for the jobs they will obtain, not just for doing research.

Far too many new Ph.D.'s are well prepared to continue a research program but are not prepared to make important contributions to other aspects of the typical college professor's job. The number of jobs, especially tenure-track jobs, that exist in doctoral mathematics departments (let alone Group I institutions) is far less than the number of new Ph.D.'s who are primarily educated for those jobs. Department chairs from bachelor's and master's departments and the Project NExT fellows in the focus groups often criticized the preparation of new Ph.D.'s to be teachers or their readiness for jobs at liberal arts institutions or other institutions where research is a relatively small part of their professional duties. Whether through an organized program such as Preparing Future Faculty, mentioned in Chapter 3, or by individual department action, it is important to rethink graduate education and to be certain that students are broadly educated so that they are prepared for the jobs they will likely hold. Broadening the education of graduate students could include topics as general as developing communication skills and learning to teach diverse groups of students, or topics as specific as offering advice on job hunting, including the preparation of an application or conducting trial interviews.

Over the past decade as new Ph.D.'s have struggled with a very tight job market, increased attention has been given to the apparent gap between the jobs for which new Ph.D.'s are prepared and the jobs that exist. As noted in the previous chapter, increasing numbers of mathematics Ph.D.'s are finding non-academic employment. The NSF VIGRE initiative is encouraging departments to prepare doctoral students for careers in business and industry.

Departments also may want to consider developing a professionally oriented master's program. Master's programs in financial mathematics and in industrial mathematics have attracted substantial attention. The AMS and MER (the Mathematicians and Education Reform Network) held a workshop titled Exploring Options in Graduate Education which pursued this issue. The AMS and SIAM (the Society for Applied and Industrial Mathematics) also sponsor a joint project on non-academic employment, which should provide valuable information to departments interested in an industrial master's program. In addition, the SIAM "Report on Mathematics in Industry" is a valuable resource for departments interested in educating doctoral students for nonacademic employment.

5. Support diversity.

We cannot argue the centrality of mathematics on campus without recognizing that historically mathematics has played a gatekeeper role, disproportionately restricting access of women and minorities to careers in mathematics, science, and engineering. This is a situation we must change.

Mathematics departments have much to gain if they assume a leadership role in creating opportunities for women and minorities at every level, from outreach programs that seek to strengthen our public school system to hiring practices in our departments. Part III of this book has a number of examples where mathematics departments have taken a lead in creating an environment that enables women and underrepresented minorities to be more successful in learning mathematics.

6. Build strong relationships on campus. Faculty should make building strong relations with other departments and the campus administration a conscious department goal.

Building strong relationships with other faculty and departments on campus is an important component of the overall goal of being a model campus department. Many deans told our Task Force in focus group discussions that their mathematics departments were too insular in their view and in the view of other campus departments. The dangers of poor relations with other departments are obvious. From time to time other departments may be tempted to teach mathematics to their own students or to send their students elsewhere for this instruction. New engineering accreditation guidelines may tempt some engineering faculty to propose teaching their own calculus to engineering students. Among the many reasons why departments cannot afford poor relations with administrators is that they are under budgetary pressure to find cheaper ways to meet large-enrollment freshman courses, and mathematics could easily become their target.

The department leadership (chair, vice-chair, and other senior faculty) should consciously cultivate campus contacts, especially with faculty from key departments who send large numbers of students to mathematics classes. These contacts may be developed through conversations at meetings of department chairs, through joint research projects, or through working together on campus initiatives. Other contacts are established when mathematics faculty are seen as good campus citizens, visibly involved in university service. Even social events can contribute to developing friends and colleagues across the campus. Along with these informal contacts, it is still important that department leaders (e.g., the chair or the undergraduate program director) regularly make formal visits to their counterparts in key departments to seek feedback on their teaching and explore possible areas of cooperation—in new campus instructional initiatives, in joint outreach, etc.

Developing good working relations with the campus administration as well as mutual respect lays a foundation for the department to influence decisions that may sustain or enhance its research and teaching program. The chair must lead this effort by communicating how the department advances the university's mission and how the department effectively uses current resources as well as explaining how the department would use additional resources. When differences arise, deans will usually listen to a chair they respect and trust.

7. Invest in strong leadership.

Discussions with department chairs and with deans and our site visits convinced our Task Force that strong department leadership is a key to building and maintaining an outstanding department. While other models may work for certain departments, strong department leadership (particularly a strong chair) can lead the department through a process of rethinking its mission and provides an accountability that assures the university administration that resources invested in the department will be used effectively. While our Task Force learned of some situations where a department suffered from the inability to get rid of an ineffective department chair, a far more common experience was that of a capable faculty member who spent the first two years as chair learning how decisions are made in the university and how to influence those decisions, only to leave the chair position after the third year.

As we have mentioned before, many departments appear focused on intradepartmental concerns and a desire to prevent one part of a department from gaining an advantage over another part of the department. All too often, a department suffers far more from the inability of the department leadership to make the case to the university administration for the resources the department needs to accomplish its mission.

Departments are well advised to seek a capable faculty member and give that person the necessary authority to be a strong chair, and then to support and value highly that faculty member's work as chair. This support should continue as long as the chair continues to be an effective leader on behalf of the department.

The leadership of the department's senior faculty is very important in establishing the goals and priorities of the department. Our Task Force learned of a number of situations where a department's ability to broaden its mission and make a significant commitment to high-quality undergraduate instruction was the result of distinguished research scholars who lent their moral support for the department's commitment to educational issues while having limited involvement in these activities.

Beyond the position of chair, most doctoral departments are large enough to need a strong, capable leadership team. The most obvious positions include a vice-chair for the undergraduate program, the graduate chair and the chief undergraduate advisor. Having senior, highly effective people in these positions is of critical importance to a department. Beyond administering and overseeing essential functions, they share with the chair the responsibility for representing the department in various forums to client departments and the administration. Depending upon the size of the department and the organizational plan, other positions may also be quite important. Such positions are also excellent training grounds for the next department chair. It is important for the department to have a capable team that works together effectively for the good of the department.

Our Task Force also noted a strong correlation between particularly successful educational initiatives (e.g., an actuarial science program, an emerging scholars program for minorities, outreach programs that work with the public schools) and the presence of a single dedicated leader who had created the program. This emphasizes the importance of identifying the right person to lead a department initiative and giving that person the support needed to create a successful program.

8. Individualize faculty workloads.

By far the best model for a faculty member is that of a teacher/scholar who is intellectually curious about teaching and is dedicated to good teaching while maintaining a strong research program. Such faculty make important contributions to the department's research mission, contribute to the graduate program as Ph.D. thesis advisors, and earn praise for the quality of their teaching. Over time they make numerous contributions to the educational mission of the department through involvement in curriculum renewal projects, the supervision of undergraduate research experiences, and various outreach activities. Over the course of a long career, the activities that attract their interest may change, but they can be expected to make regular, important contributions to different parts of the department's mission.

At the same time, it is clear that many faculty find they are much better at one aspect of the department's work than another. Over the course of a long career, faculty who were once quality contributors to the department's research mission may find that research no longer holds the same interest for them or that the quality or quantity of their research has diminished. Some faculty enjoy working with and advising students, while other faculty only grudgingly perform these tasks.

A good department chair will find a way to maximize the contributions of each faculty member. By finding work that is important to the department and which stimulates the faculty member to work hard and make valuable contributions, the department chair is accomplishing the goal of creating an environment that "encourages, enhances and enables the creative work of the faculty." This effort is hampered by a department whose approach is to insist on the same general job description for all faculty, and creates distinctions over time by rewarding some faculty with significant salaries while engaging in near punitive behavior toward faculty who are unable to thrive within a narrowly defined model of a teacher/scholar. This leads to disgruntled faculty who make very little contribution to the work of the department.

A far better idea is to match faculty with jobs that each can do well and that the department will value. Finding such matches requires considerable effort by the chair and the rest of the department leadership. This effort will likely involve a number of possible matches that do not work out, but with patience almost every faculty member can be helped to find a satisfying niche. Below, the Task Force offers several recommendations to the AMS for helping chairs with this and other difficult leadership responsibilities.

If all faculty are rewarded fairly based on their contributions, then the total accomplishments of the department are enhanced and each individual faculty member enjoys a higher-quality work experience.

9. Expand the reward system.

One of the central issues that must be addressed by faculty, especially department leaders, in doctoral departments is the question of whether the reward system hinders or enables a department's efforts to broaden their mission and establish a proper balance between the research and doctoral program and undergraduate teaching and related educational activities. The conclusion drawn by our Task Force is that the reward system is often a barrier to obtaining important contributions to all parts of a department's mission.

Our recommendation to doctoral departments is that the reward system should be guided by the following principles:

- The standard for tenure in a research department should include the expectation that those who are granted tenure have research achievements that constitute a high-quality body of scholarly work and the expectation that they have a demonstrated commitment to teaching at an appropriate level of excellence.
- The most talented researchers should enjoy the support of the university, including opportunity, resources, salary, and rank, much as they are supported at present.
- It must be easier for senior mathematicians to assume a leadership role in responding to many of the other obligations facing the department or the profession, and they must be able to do this with dignity, respect, and reward.
- There should be clear standards of excellence for those whose greatest achievements are in teaching or other educational activities, and faculty who meet those standards should share in faculty rewards, both financially and through promotion in rank.

Currently, for many departments, research achievements are the standard for receiving honor, salary, or promotion. This can result in faculty publishing mediocre research or in unproductive and disengaged faculty at a time when the department has important needs going unmet. Faculty will not spend time on activities that go unrewarded.

For a chair to carry out the preceding recommendation for engaging faculty, the reward system must recognize the full array of ways faculty can make important contributions to the department's mission. A department must determine what work is important to the department and must reward that work.

Recommendations for the AMS

Since its founding, the AMS has enjoyed a well-deserved reputation as the primary professional society for research mathematicians in America. As a result, it is uniquely qualified to provide assistance to doctoral mathematics departments as they struggle to respond to a broader mission and increased expectations from their universities and to determine the proper balance between research and education. The AMS should help these departments turn these challenges into opportunities to obtain additional resources to accomplish their expanded missions.

Our Task Force recognizes that this is not a task that can be accomplished by issuing the right report or set of recommendations. It is not something that responds to a one-time fix. Instead, it must become an ongoing activity that offers assistance to departments as they address the changing environment described in this book. We have gained an increased appreciation for the importance of giving department chairs the opportunity to interact with their peers on a regular basis and for the support that one chair can provide to another. Because departmental leadership will change regularly, there will be a continuing need to provide new chairs with the opportunity to learn about the many day-to-day responsibilities (hiring, tenure and promotion, planning, dealing with university administrators, etc.) that impact the quality of their work and the success of their department.

The Task Force recommendations supplement the annual symposium for department chairs that is held in Washington, D.C., under the leadership of the Board on the Mathematical Sciences (BMS). This symposium offers department chairs from all types of mathematical sciences departments the opportunity to increase their awareness of the major issues facing the discipline, as well as the opportunity to interact with the various funding agencies which support research and education in the mathematical sciences. There is no need to duplicate or compete with this opportunity for department chairs. At the same time, it is our judgment that department chairs at research universities need and want additional services to help them perform their job. We offer the following recommendations to the AMS that we believe, over time, can assist department chairs and their departments in responding to the broad array of challenges that impact the success of a mathematics department.

1. Continue the focus group discussions begun by the Task Force on Excellence.

Our Task Force on Excellence in Mathematics Scholarship conducted fourteen focus group discussions, including nine with chairs of doctoral mathematics departments, one with chairs from liberal arts colleges, one with Project NExT fellows, and three with deans from research universities. While the original intent in scheduling the focus groups was to gain information for the benefit of the Task Force, it quickly became clear that the discussions were extremely beneficial to department chairs. In all, the nine focus groups for chairs of doctoral departments attracted participants from 76 different departments; 32 of the departments were represented in two or more focus groups. Participation rates were highest from Group I and II public universities. Quite possibly these are the department chairs who face the widest range of issues in leading their department and who benefit most from the opportunity to discuss common issues with other department chairs.

The Task Force offers this recommendation to the AMS Committee on the Profession, with the suggestion that there should be focus group discussions for department chairs at each AMS Annual Meeting.

2. Conduct a workshop for new department chairs each year at the Annual Meetings of the AMS/MAA.

At the 1998 Annual Meeting in Baltimore, the AMS conducted a 1½-day workshop for new department chairs focusing on issues such as tenure, planning, and working effectively with your dean. The workshop was led by three successful chairs of doctoral mathematics departments, including two who are members of the Task Force. The workshop was well received by the participants, and a second workshop was conducted at the 1999 Annual Meeting in San Antonio. We recommend that the AMS continue offering this workshop for 15–25 new department chairs each year. While the focus of the workshop would be from the point of view of the chair of a doctoral department, the workshop should be open to any chair who finds it beneficial.

3. Organize a resource group of experienced department chairs to serve as consultants for departments that seek a self-assessment.

Our Task Force envisions opportunities where a department may want to take stock of what it is accomplishing and consider changes in some aspect of its work. Just as we recommended that departments invest in strong leadership, it is important that the AMS connect individual chairs with others who have experience in leading similar departments. This might happen on the occasion of the appointment of a new chair who wants to determine an agenda for the time he or she will serve as chair, or it may be a response to a regular program review mandated by the university. The department could arrange for one or two members of the resource group to visit the campus, meet with various groups within the department, and assist the department as it reviews its priorities and its goals for the next few years.

An AMS committee (e.g., the Committee on the Profession) could maintain a list of experienced department chairs willing to visit other campuses and serve as consultants.

4. The AMS should initiate expanded data services for doctoral departments.

Every five years the Conference Board of the Mathematical Sciences (CBMS) produces a significant data report on conditions in the mathematical sciences, and twice a year the AMS-IMS-MAA Data Committee produces its Annual Survey (first and second report), which is published in the *Notices* of the AMS. Taken together, this provides a rich resource of information about conditions in the mathematical sciences, including a survey of new doctoral recipients, faculty characteristics, enrollment profile, etc. It might be argued that few disciplines have comparable information about their profession.

At the same time, any discussion with department chairs eventually turns to their need for information they believe is not available but would be beneficial in making decisions and in seeking resources needed by their departments. Departments are particularly interested in data about institutions they consider most comparable to their own. For example, our fuller analysis of the Annual Report reveals that less than 4 percent of the mathematics instruction in Group I Public Universities is at the graduate level, while the corresponding percent for the top Group I Private Universities is over 10 percent. At the other end of the instructional spectrum, remedial instruction is virtually nonexistent in Group I Private Universities but constitutes about 9 percent of the instruction in Group I Public Universities.

One suggestion is that the AMS create an ongoing cohort study for departments using a selected sample of departments from each cohort to provide more complete data each year. This may require providing incentives to those departments involved in order to entice them to do the substantial work required for such data collection. A more refined cohort study, however, would be extremely useful for departments in analyzing their own situations and in making comparisons.

Eventually, the AMS might initiate a voluntary data-sharing project for doctoral mathematics departments. Participating departments would be able to access the database and seek data at comparable institutions on a wide range of topics, such as the mix of instruction (tenure-track faculty, postdocs, visitors, lecturers and graduate students), teaching loads, information on external funding, publication information, etc. The Data Committee would need to determine the criteria for participating, how information would be collected, what kind of information could be requested from the database, and what information could be released about specific institutions as opposed to other information that might only be given for groups of departments.

Getting Started

Some faculty who read these recommendations may conclude that their department has already dealt effectively with most of the issues discussed and that their department has already positioned itself with a balance between research and education appropriate to their institution's mission. Other faculty may read this book and conclude that it is important to begin discussions to reassess the department's mission but at the same time are uncertain how to begin.

The points of view suggested in this book will require a fundamental change in culture for some departments. Faculty must come to value their department's educational work as well as their research achievements. One important step is to see the difference between something being the responsibility of each and every member of the department and being the responsibility of the collective department. Aside from the basic expectation that every faculty member be involved in some form of scholarly work and that every faculty member endeavors to be an outstanding teacher, there is no role that becomes everyone's responsibility. Just as no faculty member would consider it everyone's responsibility to conduct research in algebra or topology or applied mathematics, no one should expect every faculty member to become involved in calculus reform or teacher preparation or summer programs for middle school students.

The challenge is for the department leadership to lead a process that determines an appropriate mix of roles for the department and helps faculty decide which activities are appropriate for them. As indicated earlier, faculty who are making the most significant contributions to the department's research or graduate program may not need to have any role in new educational initiatives except for giving honor and respect to those who lead in these areas. It is particularly important for the department leadership to work to avoid a conflict between research and teaching. Toward that end, we offer the following advice:

- Meet an institutional need.
- Promote change gradually.
- Make a renewed commitment to the research program.

Much of the discussion on recommendations in this chapter has centered around understanding the priorities of your university, working to help shape those priorities, and then making sure that the mathematics department is making a significant contribution to the university's highest priorities. It is with that in mind that we once again stress the importance of meeting an institutional need. If you do, then you should have reasonable expectations that the university will provide the resources needed to accomplish the work of the department.

Few people actually welcome change, especially if it involves change that they do not fully understand or that causes concern for their own welfare. A gradual approach to change offers everyone in the department the opportunity to understand the relationship between excellence in mathematics scholarship, the overall health of the department, and the overall health of their institution. It is also important to pay particular attention to the department's research mission, to make certain that it has strong support from the departmental leadership during any period of time when the focus is on instructional issues. Attention to this issue can help avoid a conflict between those in the department most interested in protecting the department's research mission and those focused on expanding the department's commitment to educational work.

The remainder of this book contains additional information that we believe will be useful. First we offer readers an opportunity to listen to the mathematics community as they expressed themselves in our focus groups. We then take an in-depth look at the educational activities of five departments we visited and report on a number of other examples that came to our attention. Finally we offer a number of thoughtful essays from leaders in the profession and provide a number of resources we believe will be of benefit.

Part II

Messages

Chapter 5
Comments from Chairs of Doctoral Departments

For over two years, groups of department chairs (and ex-chairs) from mathematics departments granting Ph.D. degrees met in focus groups to talk for two or three hours at a time. In a real sense these focus groups were the central accomplishment of the Task Force. They gave almost one hundred chairs of doctoral departments a chance to share experiences, to ask questions, and to offer advice. For many, the experience was both a relief and a revelation.

What did we learn in the focus groups? What advice did all those chairs have to offer? The first question is easy to answer: We learned about the problems, the concerns, and the successes of mathematics departments across the country. It was informal learning, the kind of knowledge one gains in casual discussion, but it helped chairs understand their own situations in the context of the broader mathematics community. The second question is much harder to answer, and this is the central dilemma faced by the Task Force.

We have included below samples of the many notes taken during the nine focus groups, collected into categories to show the common threads present in almost every discussion. These are indirect quotes, extracted from discussions over a two-year period.

Many Departments

Almost every participant told a story about a department: the pressures felt, the special problems in a particular university, the way in which the department grew or contracted. The central lesson of all these stories is simple: While departments face many common problems, they also differ in essential ways.

In reading the comments, however, some general trends become apparent. In a typical department the number of faculty is decreasing, majors are decreasing, the graduate program is shrinking. There are substantial decreases in upper-level undergraduate courses. The only things that seem to be increasing are the number of non-tenure-track faculty (either postdocs or part-time) and students in remedial courses. The feeling of many chairs was expressed by one person's comment that there are "too few faculty and too many students."

Many of these trends are borne out by the surveys (for example, the most recent CBMS survey, see Chapter 21), but the surveys as well as the comments of a few chairs show that the situation is neither simple nor uniform. The number of tenure-track faculty is indeed decreasing at doctoral-granting universities, but to a large extent this decrease is offset by an increase in non-tenure-track positions (mainly postdocs). In the meantime, there has been a dramatic decrease in enrollments at these institutions. One might wonder why the situation isn't improving. The anecdotes here, along with some more refined analysis of the data, show why it is not.

Although both enrollments and faculty are decreasing in general, these trends are not uniform. Calculus enrollment in Group I private universities, for example, decreased from 1992 to 1997, but it increased in Group I public universities. In the same period, graduate student enrollments were reversed in these two groups. Many of the comments in the focus groups reinforced the vision of departments adjusting to such changes.

However, mixed with the problems were stories of success: departments that faced the future with optimism because they had found ways to secure resources or respect from the university. In differing degrees, many departments reported success even when they despaired at the problems.

There were other variations in departments. Some had many part-time graduate students (who are mainly older). Some were applied departments with no responsibilities for calculus. A few indicated a tradition of emphasizing teaching, which they felt gave them an advantage in today's climate. No matter how they varied, however, every department appeared to be under pressure.

Instruction

What are the pressures felt by departments? Over and over in the focus groups it became clear that the answer was overwhelmingly instruction: how to improve, how to create a better image, how to convince faculty to undertake time-consuming projects. The focus groups vacillated between bemoaning the sorry state of students ("they take no responsibility for learning") and castigating the community for its lack of effort ("math departments do a lousy job").

Nearly every group of chairs talked about the need for smaller classes, and many expressed an interest in gathering evidence that small classes are better. Many departments had already achieved smaller classes or planned to do so soon.

A number of chairs worried about the evaluation of teaching: whether it is good or bad and the increasing pressure to carry out a more elaborate evaluation process. There was some grumbling about treating students as customers, and the comment was made frequently that good evaluations are not necessarily correlated with good teaching. But there was also the recognition by many that mathematics is under pressure to be more accountable.

What are the major changes in instruction? Chairs most often mentioned computers (or calculators) in the classroom. Some commented with pride that their department has computer labs for the students, nearly always followed by the comment that such labs require substantial resources. Many indicated that graphing calculators seem to be a more practical alternative.

Calculus reform was a topic brought up in every focus group, often by chairs who were apologetic that reform seemed to have passed them by. For those who indicated that their departments were engaged in reform, most talked about computers first and the particular reform text second. Group learning and other experimentation was mentioned less frequently. In many departments, reform appeared to be done in only some classes, and the problems of integrating several versions of calculus were mentioned by a number of chairs. It was clear that nationwide reform efforts had made most departments consider instructional issues, even when they were not actively engaged in reform themselves.

The instructional problem most often mentioned by chairs was college algebra, which often led to a discussion of remedial courses taught by the department. Most departments seem to accept the necessity of providing large numbers of remedial courses. Some, however, have alternative programs, either directing students into slower-paced university courses or sending them to other institutions for remediation. Only a few indicated that they were able to raise standards for admission.

The number of mathematics majors in departments was seen as closely connected to their instructional program. Overall the number of majors is decreasing, and surveys indicate that between 1992 and 1997 the number of majors has decreased by about 12 percent. That figure hides an important difference between universities, however. In Group I schools, the drop has been slightly above 20 percent; in Group II and III schools it has been closer to 5 percent. That difference was reflected in the focus groups by comments from some departments that the number of majors had actually increased in recent years. A few provided some details about programs designed to attract and to retain new majors.

Many fewer chairs mentioned problems (or successes) with teacher-training programs. Indeed, when teacher education was mentioned, it was often viewed as an outreach activity rather than as an integral part of the instructional program.

Graduate Program

Graduate programs were discussed in every focus group, and a number of departments reported downsizing their programs. This downsizing was sometimes mandated. The comments about the effect on the department were muted, however. Considering recent data, this understatement is remarkable. Surveys suggest that between 1991 and 1997, the number of full-time graduate students at Group I, II, and III universities dropped by about 20 percent. The number of first-year graduate students during the same period decreased by about 28 percent. These are dramatic decreases. But even this hides a more remarkable difference. The decrease in first-year graduate students from 1991 to 1997 for Group I universities was nearly 40 percent, while it was only 12 percent for Group II.

There were few remarks about the effects of these dramatic shifts on departments, except indirectly. It was clear that some departments are considering reinvigorating (or creating) master's programs for their students. A number of departments talked about industrial components for graduate degrees. A number of others indicated extra efforts to train graduate students as teachers, both for their work during graduate training and for their careers later. All these things are designed to provide a better graduate program, which ultimately will attract better (and more) students.

It was remarkable that almost no chair indicated that his department was considering substantial modifications to their current doctoral program, except for adding an industrial component and teaching experience. There was virtually no discussion of curriculum, qualifying exams, or dissertation requirements. Only one or two chairs knew what the average time-to-degree was for their graduate students. Few departments tracked students after they left.

Deans

Almost every chair recognized that deans require evidence in order to be persuaded to give more resources. Often, however, chairs were unclear about what evidence to provide. There was a recurring theme in focus groups: a call to committee to provide a report that any chair could take to the dean, providing convincing evidence that the dean should give mathematics more money. This was the "magic bullet" approach towards dealing with the administration, and it was extremely appealing to everyone, including members of the Task Force.

Deans (and other administrators) were viewed by chairs with a gentle antagonism. A few commented that members of the administration held nonacademic values, but most believed that deans were other academics who merely needed more knowledge about the mathematics department and the benefit a healthy department would bring to the university.

Chairs of departments that had been successful in securing major resources from the university provided occasional advice (clearly indicate the intended use, build support from other departments, etc.). In many cases, however, these successful chairs indicated that there had been special circumstances that permitted them to compete for resources. There was some good advice, but there were no magic bullets.

Some Advice

There were other issues brought up occasionally at the focus groups: some interest in development and fundraising, as well as a general concern about library budgets and the future of subscriptions. Remarkably, however, there were few topics that compared to the discussion of instruction. It was a constant theme for every chair at every meeting.

Some chairs offered good advice:

- One of the problems of mathematics is that mathematics is invisible in the political structure of the institution.
- Mathematics does a poor job of selling itself. Our initial courses should provide a good experience for students.
- We need to show that "math is a smart major" and that math majors make more than other science majors.
- The major problem is communication between the math department and other departments. The other departments need to understand the pressures on a math department; we need to make an effort to go out to the client departments to get information and feedback.
- Mathematics is really key to what is happening in the institution. As a discipline mathematics must do more thinking than anyone else about the way it educates its students (at all levels).

- Technology is an enhancement, not a replacement.

The topics that did not arise are almost as informative as those that did. Except for one or two incidental remarks, almost no chair talked about a perceived need to reinvigorate a research program in the department. A few mentioned the problem of faculty who were unproductive in research, but only in connection with variable teaching loads. Few talked about interdisciplinary programs outside the context of instruction, and none mentioned interdisciplinary research groups. The insularity of departments was never mentioned, not once (see Chapter 6). Almost no one mentioned conflicts between pure and applied mathematics (see Chapter 6). No one mentioned an effort to change in fundamental ways the basic doctoral program: distribution requirements, qualifying exams, dissertation.

Mainly, chairs were melancholy about the prospects for mathematics. Mixed with their pessimism, however, was a belief that mathematics is important to the university and to the students. Except to point out that many administrators have someone in their family who had a bad experience with mathematics, there was an inability to explain why we are unable to explain the poor reputation for mathematics.

Chairs understand that there is a problem; they do not understand its nature, its scope, or its solution.

Comments from Chairs

Stories of Departments

♦ We have 32 tenure-track research positions, down 2 since 1990. The university has increased in size every year, and the state's projection is that the university classes will skyrocket. In addition to the tenure-track positions, we have 10 budgeted instructor positions created in the 1970s and 1980s for a very large precalculus program, started as a remedial mathematics project, and we give credit for these. We changed the way we taught calculus; it is taught by faculty and graduate students. The only support we've received for any of this has been two extra graduate students.

♦ There are too few faculty and too many students. The department has received additional funds to hire part-time faculty, but we mostly have no new faculty lines. One exception is a new line to develop applied calculus and precalculus. The department is hiring year-by-year full-time calculus teachers with higher teaching loads. People with master's degrees are hired to teach twelve hours.

♦ Our university got the downsizing bug. Essentially there have been no new people coming into the department. This seems to be happening in a lot of universities. We teach fewer classes, and of course classes get larger. Over the past fifteen years we have seen a tremendous drop in interest in mathematics among American students. We have a large number of Chinese and Russian undergraduates. We can still give courses to maybe 10 students.

♦ We have a number of tenure contract faculty, who teach far more students than regular faculty members and more sections. We are experiencing a larger and larger precalculus burden — very dramatic and significant. None of the regular faculty interacts with these courses. We are looking to make one appointment to replace three retiring faculty. We have two faculty who have indicated they will resign. If you look at the total teaching load of the department, graduate students do a large amount of remedial teaching, and the regular faculty have only about 50 percent of the contact hours for the rest. The regular faculty teach the graduate courses; the advanced undergraduate courses are becoming smaller and smaller. I am trying to improve the latter, since a mathematics major demonstrates significant thinking skill and will improve your ability to be hired.

♦ We just went through a large number of retirements: of 17 faculty, 6 took early retirement with eight years' service credits. We recruited for 2 faculty, and for many years we have had temporary faculty. One-third of our calculus is taught by permanent lecturers who eventually are made full time. We also use itinerant lecturers, but we need a critical mass of people in order to be viewed as a research university. The administration is most concerned about costs per course.

♦ We have lost a lot of the senior faculty members, having gone from 70 to 50 senior faculty. The number of junior faculty that we are allowed to hire is dropping very fast; grants for national needs no longer exist. Our computer labs are being closed and/or are threatened. It is hard to staff courses, since they try to insist that junior- and senior-level courses have at least 35 students in the classroom.

♦ We have a new president, and he came in with a big problem of recruitment and retention. He identified college algebra as one of our biggest problems, and he put in place a policy of no rookies on rookies in the classrooms. We have four new faculty positions, and we have gotten extremely positive responses on the campus. We have had a tremendous shift in recent years, and almost all the departments require these elementary courses. We are running faculty workshops in the summer, and for one week the faculty have to learn how to teach these classes. Faculty are preparing the material, and the first one is being done in summer of 1997. The workshops will be developed by the people that are teaching the courses.

♦ There have been lots of small problems in the last ten years, and several big mistakes. The business school created a quantitative reasoning department, and the math department was opposed. All this might have worked, but they started to look more and more like a math department, and they are now teaching calculus and engineering. We are trying to come up with some resolution for our problems. We need to resolve all these applied-math-versus-math issues.

♦ Our department is less traumatized by recent events because we have been a department with a strong commitment to teaching. All the courses above precalculus are taught by regular faculty, and it has stood us in good stead. For us, teaching qualifications have always been important; it is well understood that our teaching mission is a major part of our sustenance. We all teach two courses per semester. We had a period of confrontation between engineering and mathematics. They claimed that they could teach calculus equally well for their students. We made the case that the university had a strong investment in the mathematics department and that its viability depended on retaining our instructional mission. This case was won, but it resurfaces frequently, and math departments need to be vigilant about the threats that this presents. This is an important lesson for people to keep in mind. We are a traditional core mathematics department; we don't have good industrial contacts. One thing that has always struck me is that when mathematicians look at the life of a department like ours, they see something very different from administrators. There is a tremendous amount of mathematics that goes on in seminars in which we invest a large amount of time. None of these show up in our teaching loads, but they take up an incredible amount of time.

♦ Our department is nontraditional in several ways. We have a very broad-based core that resembles a kind of 1960s mathematics department. But we also require everyone to take computational mathematics and probability and statistics. We also have a required applications component, which requires of the students a semester of student work. We also have students who come back and participate in instruction. And we have a couple of on-campus internships with departments such as engineering. We have had a student in math education working in a school district on curricular matters. The students we have been graduating—every one of them—have gotten jobs. A couple of our students were made offers at the places where they had the internships. Most of our students who want academic careers are interested in teaching careers, and the application component helps them quite a bit. Internships are set up by a single person (almost like a director of graduate studies), and the hard part of the job is to keep up the contacts with industries. It's not easy to get someone to take on the job and do it with enthusiasm.

♦ Our environment is that of an urban university. We have been absorbing two to three cuts each year. We used to have 75 faculty, but the Institute must shrink by 30 faculty members. We now have 70 faculty and 100 graduate students, which is down 30 percent. We have about 6 to 10 graduate students finishing each year. The morale of beginning graduate students is very low.

♦ Our situation and programs are different from what I hear. We are a large urban comprehensive research university. We have some unique features: a large part-time graduate student body, a large master's component, including a master's in applied mathematics, and a program designed for part-time study and for people whose undergraduate degree is not mathematics. There are advantages and disadvantages. Many Ph.D. students are employed while pursuing their degree. Our programs satisfy a need for people who want intellectual stimulation; this is also true for people working in industry. This mission of a graduate program needs to be recognized. We have some ties to local industries, and we have a couple of interdisciplinary centers working on nonlinear analysis. The department has attracted good to excellent faculty, and we've seen a dramatic increase in the quality and accomplishments of students. There are new directions in areas of sciences, especially areas that are good for interdisciplinary work. Our greatest difficulty is that our course load has gone from 6,000 to 10,000 students, and this has put a severe strain on our resources. We cannot afford to experiment.

♦ The department has 32 faculty, 27 tenure-track, and 5 visiting/temporary positions in research. There are 35 graduate students, 30 of whom are TA's. About 5–6 Ph.D.'s finish each year: about 3 find research jobs, and 2 teaching jobs. TA's teach fewer than 7 credit hours and handle sections of fewer than 25 students. There is a successful training program for TA's. First-year students don't teach; they're attached to a senior teaching assistant to help with tests and grading. Second-year students teach and are monitored by a committee that helps them. There is a center for teaching at the university and a lecture series on great teaching. Graduate students are provided with small group analysis and discussion. Lectures are videotaped. Similar help is given to new faculty. Since good teaching is expected, graduate students do not have difficulty with these expectations. There is a $500 award each year to a TA for excellence in teaching. Since good teaching is expected of faculty at all times, at promotion, faculty can be judged on their research.

♦ The mathematics program is only about twenty-five years old. There has been recent work on the graduate program, and there are now 85 full-time TA's. The department graduates about 12 Ph.D.'s each year, all of whom (except those with personal reasons) get jobs. Most get academic jobs at both teaching and research institutions, and about 1 or 2 go into business per year.

♦ There are 35 Ph.D. students, with 3 or 4 finishing per year. Given the present job market, more will not be supported. A large number of students work outside part time, and courses must be adapted to work schedules. Students come with very wide backgrounds and preparation. Large numbers of students begin at levels below calculus. The university is held to a standard set by the state and is underfunded. Large numbers of part-time faculty are teaching. The department must live with its tight budget. Researchers teach a 2-2 load and others a 3-3.

♦ We are organized differently from most departments. There is a core math department and an umbrella math department that picks up most other things. The two departments cooperate in ways that are productive. We have programs that capture traditional math majors, and then we have the professional terminal math major for a master's. This is a first step: many students are not ready to go into a traditional math major, and they enjoy something more applied. For them we have a professional B.S. program in math with no proofs. In many ways, this is equivalent to an engineering degree.

♦ We have 10 tenure-line faculty. Our graduate program is dominated by service; we are essentially a technical university. This means we get lots of students in our courses. That's bad because we have large financial pressures, and we're forced to put much more effort into service teaching and don't have much left for research. We have reduced the size of calculus classes, have introduced Maple, and switched to the Harvard curriculum, all this due to the efforts of one faculty member without funding.

♦ We have 27 tenure-track faculty. We have some new instructorships, funded by a former member of our department. Two of them are labeled research, and one is labeled teaching with a scholarly agenda. These instructorships have a reduced teaching load so they can do curriculum development and research. This effort has just started, so there is still room for growth.

♦ We have 4 semipermanent instructors. The regular FTE's have stayed flat for the last few years. We are down to 26 FTE's after some people retired; there is only temporary money. The deans look at this temporary money as the only thing that is flexible in their budget. You should know that the budget in the math department is determined by the number of FTE's, so with the retired faculty our budget has gone way down. Yet we must still teach the same number of students. We are under pressure to increase class size. We are trying a calculus class of over 300 for the first time.

♦ We have math and computer science in the same department. There was no computer science department, and in 1976 we built a computer program within mathematics. We give the only Ph.D. in computer science in the state. We created an institute of computational mathematics to take advantage of the applied mathematics in our department. We share research colloquia with physics and computer science. We have about 65 supported graduate students, one half are in computer science.

♦ We have variable teaching for research and scholarship reasons. We have a significantly sized department with full professors, and some have slacked off in their research. Every math course we teach is no larger than 35 students. We decided to look at the faculty we would like to give reduced loads to and then make changes. We have reallocated our loads to help the research efforts.

♦ We lost some faculty to retirement, and we have had to concentrate our research in fewer areas: applied math, differential equations, and analysis. We have been forced to limit the areas. We have not decreased the program; we have simply become more efficient.

♦ We started out as a university with very few pretensions. There is a clash of values, and this varies from department to department. We have a large number of positions at a time when the market is not good. We have added 15 positions since early 1991. Our department has undergone a number of changes. Because we have responsibility for center management, and the math department is usually the cash cow, we are in a good position. When the center management model came into effect, they said that our income would be determined by the number of students and our overhead on grants. In theory, 72 percent of the money comes back to the department, but of course the real amount still depends on the dean.

♦ Applied mathematics has different instructional demands: we teach no calculus; we have large undergraduate courses in differential equations; and we serve students across the sciences, especially in engineering. Applied mathematics emphasizes graduate programs: 60 percent of our enrollment is at the graduate level. We have a number of undergraduate majors in applied math (operations research, statistics, applied math, and economics, regarded as a great path to business school). We also have a very large visitor program, which is related to the mission of the department. The visitors outnumber the faculty in some years. The university does not contribute to this; it is done entirely from grants. The structure is good for both departments, and there are joint appointments. But our department is very different from a typical mathematics department.

Instruction

♦ Our university has been retrenching for several years. Research support has been time honored on campus, and there is a feeling that if you show an interest in teaching, you are not interested in research. We need to encourage and to acknowledge teaching, and we need to try to say that teaching does make a difference. There is a problem in balancing teaching and research. If NSF had more programs supporting teaching and the K–12 interaction, then you could say, "look, this person has a grant." We started a young scholars program for minority students in the state, hoping that national groups and companies would continue. This did not happen, and they decided it was too expensive to turn it into a recruitment device. It now is more remedial than academic.

♦ I have found that we don't instill in our students a responsibility for learning. The attitude of our students has really deteriorated, and they are not very responsible. This is a part of the equation that is never mentioned. We have to figure out how to do this. George Cobb at Holyoke has pointed out that there has been a great effort by faculty members to make students happy. Students and faculty are supposed to be working together, using the methods of TQM to get the students to do their work. That's the problem.

♦ The committee had better go beyond proving that calculus is better taught in small sections. Proof that it's better in small sections isn't enough; we need to know how to implement smaller sections.

♦ We have been very good at denigrating ourselves publicly and privately. We also need to show off some of the good things we do. We have not done enough to show off some of our successes, the ways in which we have improved our teaching. When we ask other departments how we can best help them, the focus tends to be on the negative.

♦ To be truthful, most math departments do a lousy job. I have looked at it from the point of view of the engineers as well as of the mathematicians, and I find that most mathematicians have no idea how to create a syllabus for a course. They have no idea how to prepare the kids and make sure that they can do homework on a regular basis.

Stand at the back of a lecture and take a look around the room; you may realize that the majority of the people are not paying attention. Mathematicians are not doing their job.

♦ We emphasize good teaching. We have always had a program in which we emphasize good teaching. We give awards to the best undergraduate teachers, and we show in many ways that we recognize good teaching.

♦ We teach many service courses in which we are teaching people who are not happy about being there. We need to educate people, to teach those who don't want to learn what we are teaching. For that reason, it is important to teach communication skills to teachers, promoting more active learning.

♦ Are the same people who are asking for assessment also asking you to measure student learning? There is not a perfect correlation between attitude and learning. You cannot infer that if students feel good about their experience, then they have necessarily learned more.

♦ One reason people (including me) are so quick to criticize mathematics is very simple: Anyone in engineering or the hard sciences will tell you that without a good mathematics program they are in trouble. I am very critical because some faculty think it is not difficult to teach. It is hard work, and it takes lots of preparation, regardless of our training. Think about how it ties in with other disciplines and what students need to know. I have realized that I ought to teach the people in front of me the way I would like to have my own children taught.

♦ The real issue is about the intellectual case, the problem we face when going to the dean to hire. The dean wants us to use anyone to teach mathematics and thinks faculty is interchangeable. Other professors would never teach a course if it isn't their research specialty.

♦ We changed from large-section to small-section calculus about twelve years ago. This required many more faculty, and the chair then worked out a very ingenious method of getting them: small college teachers would come and teach some of those sections, which represented for them a small teaching load for which they got compensation while they were on sabbatical from their institution. They contributed a lot to the life of the department. Once the foundation support disappeared, the college picked it up, but they have become increasingly critical of the large number of visitors. In order to maintain the program we need a dozen or so faculty more than we actually have. We have met our needs in the last few years by increasing the number of other visitors.

♦ We have interdisciplinary grants at our school and a grant for interdisciplinary work with the engineering school. The math department is under a lot of pressure to become the Baskin-Robbins of calculus —to provide all the flavors. Engineering would be just as happy to hire mathematicians to work directly in the engineering department to teach mathematics.

♦ The idea of treating the students as customers has led to a lowering of standards.

♦ We have two different calculus classes, and I am concerned that the department seems to be marching in different directions. The students would like to have more technology put into use, and I think that the faculty would like it also. Freshman love chemistry and other sciences, but they don't like mathematics. The reason is that the other sciences pay more attention to their beginning students. We have a placement exam — pretty boring. Their chemistry homework is hard, but they spend a lot of time doing it anyway.

♦ There is a tension between undergraduate teaching and graduate training. We must balance the need to cover courses, to have assistants for teaching, and to mentor our graduate students. An administrative survey showed that math scored lowest on the number of undergraduates seen by a typical faculty member per year.

♦ Americans are really in the minority at my school; the problem I find is not the language but a culture gap. A prime example is an eminent mathematician I recently appointed. I was actually blocked when I first tried to appoint him. I promised to teach his courses if this person failed in the classroom. On that basis he was appointed. Although he has a tremendous problem with his English, he was nominated for a teaching award.

♦ It is assumed that if someone gets good teaching evaluations that they are good teachers. I would like to have something to show that teaching evaluations by students should not be used as the only measurement of good teaching.

♦ Teaching evaluations are largely done by students, and we are now starting to do faculty evaluation of teaching. Mentoring in the last five or six years has been more successful with Ph.D. students. We give them the opportunity to teach over the summer, and this usually helps the students to find employment.

♦ The school of engineering and physics has recently threatened to teach their own version of calculus. After much discussion, mathematics finally has the course back, but only after swearing up and down that they were going to do a good job.

♦ One way to measure success is to track the performance of the teachers. It is important to look at things other than student evaluations; you need to look at the performance of the continuing students and see how successful the students are after having had a particular teacher. It is important to realize that "you cannot fatten the hog by weighing it". English departments give the impression that they are doing assessment; mathematics departments don't.

♦ Our college decided that research was going to be very important, but teaching is considered sacrosanct. Teacher evaluation is done by writing to alumni (about 100) along with department recommendations. Teaching and research are being weighed equally these days; if you are not outstanding in both, you will not be promoted.

♦ We have been using our own homegrown calculus book. The book gets very low ratings by the students, and it has affected our student evaluations. They want a real book that they can use for reference.

♦ We talk about quality, yet we don't define what we mean by quality in teaching. We also need to say what defines success and what is the success of the student relative to this cost analysis.

♦ College algebra is a problem. It would be nice to know what general math requirements are in other universities and what works with these students. A study was done to predict performance in college algebra. A combination of high school GPA, admissions test scores, and a couple other factors are used to predict student placement. This method is used, since funding for tests has not always been available.

♦ Remedial and precalculus is where we have most of the problems. The administration insists that we have faculty — professors — teaching them.

♦ The university has a quantitative reasoning graduation requirement, which increases the load in mathematics. The college algebra course is being rethought to serve better as a terminal mathematics course. Collaborative learning is being used. TA's are being prepared better for these courses.

♦ The importance of mathematics courses to other departments varies tremendously from university to university. Who in the other department are you trying to convince? We ought to listen to other people and gather information. We need to listen to what they need and then respond. But we also should remember that we are the mathematicians, and we are the ones who know how to teach mathematics.

♦ We had a series of lucky occurrences about five years ago. There was a task force on undergraduate education, and it met for two years and produced a report. We did an experiment, taking students who were predicted to be at the bottom of our calculus classes and teaching them in small sections. They performed above average on the exams. We ran a few more pilot sections, and we made a proposal to teach all entry-level calculus in small sections. We had only a little data, but it worked. The next thing that happened was that there was some reallocation, and we moved from big calculus to small calculus. The sections of 35 and under are taught by Ph.D.'s, although we have a few large sections left. We copied some of what Michigan has done, hiring people on two- or three-year appointments, not on tenure track. We have gotten rid of a lot of stuff in our curriculum, things like integration formulas. We didn't get flack from other departments because we invited them to share in the discussion when we changed. The students are clawing over each other to get into the small sections. And we give them first-come first-served classes, with the exception of one dean who has insisted that all his students attend small sections.

♦ We have what is called a college algebra course, and this course (or something higher) has to be taken by all students. We have a nonengineering calculus sequence and an engineering calculus sequence — all these are taught in small sections. They are almost all taught by part-time instructors or second-year-and-above teaching assistants. The lectures consist of two groups of 35 students. There is no placement: the students can assert their rights to take anything they want. We have successes in recruiting minority students from rural backgrounds who have demonstrated potential to excel in mathematics. They are recruited primarily due to the energy of one individual in the department. These students attend ordinary lectures and an additional six hours of recitation. They learn how to learn, and they do well. It is expensive, but these students are getting better grades consistently, and they seem to be doing okay in engineering, etc.

♦ In working with engineering departments we are told we are not using enough computers. They would like to take over these courses. We have downsized our department, and we are downsizing our graduate program even more.

♦ The college has just implemented a policy for promotion from associate to full professor, requiring documented efforts in the area of instruction. Somehow candidates need to have in their portfolio of activities something that reflects the quality of their teaching. On the other side, we work very aggressively with people who are having trouble with teacher evaluations. We "penalize" bad teachers by splitting their classes, making them smaller. That way fewer students are subjected to a bad teacher, but it is not much of a penalty.

♦ There are no changes at our university; we teach as always. Teaching is what we are there for, and classes are all small. Students get constant feedback, and faculty are always available in their offices, waiting for students to drop in. This is the liberal arts tradition. A large number of math majors go on to get Ph.D.'s. There is nothing revolutionary in our department: we have not made any changes, we have not looked for additional resources. We have tried technology: for example, computing in modern algebra. But this has not seemed to make any difference.

♦ We teach some sections using Mathematica; we need a computer lab, but this is financially burdensome. I hope that the College of Arts and Sciences will see fit to create a computer lab for us instead of the math department trying to do it alone. Regular faculty members are very reluctant to go into new approaches to calculus. All our regular calculus classes use a calculator-based class, thereby removing the burden of having a computer lab.

♦ Five years ago the university instituted a 55-credit general education program. One of the requirements of this program is a mathematics course, and we have started some new courses: math appreciation and business calculus. They both now have 400 students each semester. General education courses are supposed to be taught by tenure-track professors, and all of our classes are taught to 45 students or less, but there are no resources to help us with this. I've noticed that the majority of faculty wants to teach behind closed doors; the most time they spend is researching what they are going to teach. People need to read journals on improving their classroom techniques; we don't have seminars about teaching.

Reform

♦ In calculus reform we have done nothing, except for individuals that have done their own thing. We have a large number of faculty looking at reform from the interdisciplinary point of view. Group learning, more interactive classrooms — there have been changes in the upper division that don't involve a lot of students.

♦ The issue of what we are trying to teach in calculus has to be addressed. Should we try to teach skills? That's what the students believe we should teach. We are not able to teach beyond skills when we have to teach classes that are too large.

♦ Calculus reform has bypassed us. Most things depend on the chair. We have tried to bring in some interesting mathematics from industry, and we got NSF support to do that. We did get a grant to do theory and to do some computation — this was our one venture into education.

♦ We are doing quite a bit toward instructional reform in the calculus sequence, running some experimental sections alongside the traditional sections. Only senior faculty are teaching the experimental sections. Two are math education specialists, and three are mathematicians. They have weekly meetings on how this is all going. The final aim is to have all small sections, but this isn't possible with the number of faculty available. We need to do more in the direction of technology and with the number of math education specialists.

♦ We are not making dramatic changes, just trying to keep abreast of what is happening. We teach a lot of calculus in sections of 40, and it is taught by regular faculty. We have done sections with Maple and extra hours in a Mac lab, but we don't have the resources to do this with everyone who takes calculus. We ran Harvard consortium materials and used graphing calculators. We have had a large section of calculus with about 120 students and a common final examination of all sections. We kept statistics to compare, and the results were that the people in large sections did better. Also, those who teach in large sections tend to be better teachers and the students tend to self-select, so that those who select the large classes feel comfortable with calculus already. Those students were taught by regular professors except for the one hour a week with a TA. We teach in one large section, and the other sections are 40.

♦ We are now using the new Harvard calculus. In first-semester calculus 73 percent of the students passed calculus with a C or better, and in the next semester with the old method 53 percent of the students passed second-semester calculus with a C or better. We put an incredible amount of resources into this.

♦ We are somewhat behind here. We use Maple throughout the upper-level curriculum. We teach many sections of two large courses at the lower level, and we started introducing graphing calculators into some. The problem is that we teach those courses to seven thousand students a year. Some faculty are happy while others are not at all; it is a big job to introduce calculators. We are changing the way we teach the courses, and we will try to phase in those changes during the next two years. We are formulating a fairly radical proposal for teaching calculus in small sections and bringing down the credit hours of each course. Students have to be ready to take the next set of courses, so the courses need structure and faculty need constraints. But you have to give people freedom to teach in order to make them creative; people will be better teachers if they are creative.

♦ When we implemented the new calculus, we decided to do all the sections the same way. The reaction of the students was immediately intense and negative, since they had no other place to go. The program included using Maple and group work, mostly focused on working together in computer groups. They had trouble getting together.

♦ Three years ago we changed the general structure of our courses. We have special sections for business and honors. We also have a special section in which computers are used, although we have used things less complicated than Mathematica. We introduced graphing calculators in a weak calculus class and it didn't work, so we backed out. This fall we are going to make using calculators optional. In the fall this course is large, with about 500 students. All in all we have been very conservative. Strangely enough the engineering school has also not been interested in this kind of technology.

♦ The department is using a traditional curriculum while beginning the process of reform. The dean is pushing, while there is concern about how to bring the faculty on board.

♦ The significant change from the point of view of education has been the Calculus Reform project. The essence of our program is that we have turned calculus into a laboratory course — the ideal science course. We actually expect students to work and think more among themselves rather than with us. By the end of the freshman year we have students writing significant reports, and we have abandoned the idea that if we don't tell it to them, they will not know it.

♦ We come from a conservative background. We have a secondary school associated with the university, and they were using graphing calculators. We were told that if we didn't get our act together and do something reasonably modern, they would cease to recommend their students to us. There are faculty members who refuse to use technology. All sections of calculus are not taught with calculators, but the students are very responsive. We now have two sections, but I guess that four to five years from now almost all the teaching will be done with graphing calculators. It's largely been a positive experience.

♦ We started with students who were beginning calculus, experimenting with graphing calculators. We don't have enough computer labs to deal with high-level software in calculus, but we have started using Mathematica and have been successful. We would like to experiment more with it. That was the extent of our department's efforts; we put in a request for more resources, and while we have not changed the curriculum, we will add some special sections.

♦ We have undergraduate math counselors in the dorms providing tutorials. We have consultation rooms and about four different support systems for students. When we went to using Harvard calculus, we used a project approach. We assigned projects to students and had the students work on the projects in groups. It seems to have had a big effect on retention: the students don't disappear anymore.

♦ Five years ago we got a lab and workstations and began using Mathematica and Maple. We now have a lab fee. We got a campus site license, and all students have projects to do. We now have more small learning groups using graphing calculators and are experimenting with graphing calculators in calculus. The nature of instruction is changing dramatically: we are renovating a classroom to use two or three different kinds of computers to create a high-tech environment. This one classroom is primarily dedicated to our department.

♦ Calculus has diversified and comes in many flavors. Biology is one new version of calculus. We have outreach programs using Mathematica, and some of those programs don't require the students to come to campus. This is labor intensive, but undergraduates are used in part of the program.

♦ For six years we have been experimenting using Mathematica. Our faculty is too small to do it in many sections, but we did it in a couple of sections. We tried to offer small sections, but the students voted against it: they perceived this as being more work. We decided to do something different in order to incorporate technology into teaching, and so we decided to go with HP graphing calculators in all calculus sections. Then we decided to introduce them in precalculus courses, and the students thought this was a reasonable thing to do. They would like to see us go to laptop computers. It really is nice for all the students to have the same machine and to have it with them everywhere they go. The most complaints are from the transfer students. We are running a $40 ten-hour workshop for transfer students to get them up to speed on the graphing calculators.

♦ Our dean feels that we need to convince the other departments that reform is a good idea. We have been on the reserve system for the last ten years; the business and engineering schools have incentives to teach calculus themselves. They have their own agenda and are not worried about the best way to teach mathematics.

♦ Over the past five years there have been big changes in the attitudes of the faculty members toward the importance and involvement of student learning. We started out with involvement in the Calculus Reform project viewed as a technology thing. As we got into the project the central focus became getting more student involvement and more cooperative learning. We have instituted a large training program to train faculty in how to do this. The faculty has been very receptive, which has had an impact throughout the curriculum, and we have more people interested in getting involved in these projects than ever before. The major effect has been a shift in the way faculty members think about their job. There has been a lot of emphasis given to how students receive the courses, and in response we have been able to cut class size down to about twenty-four in 30 percent of our calculus classes.

♦ The university has good technological resources. Most upper undergraduate courses have Mathematica available. There is a state-of-the-art classroom with thirty-five workstations. The university has a PEW grant to train faculty to incorporate technology into their teaching. A large quantity of materials had to be developed to support this. These materials are available to all participants. It is difficult to use technology. Few calculus sections use computers. The cost of labs is high. One solution might be networked classrooms with students owning laptop computers.

♦ Undergraduate courses are being reformed. There is a new interdisciplinary course being written by sixteen writing teams; it will use team teaching. The issue of granting teaching credit has not yet been addressed, as this program is still being planned. The department has funds to support faculty outside mathematics in this effort, and modules are being created for science courses where mathematics is needed. Work is being done to create a more problem-oriented core for the undergraduate major. Mathematical modeling is becoming more the focus. The department is asking what it means to think rigorously and is trying to put more mathematics earlier in the curriculum. This is part of the effort under a preliminary NSF grant for undergraduate curriculum development.

♦ We have been involved in a number of reform movements. Three years ago we did calculus reform with an NSF grant, and we developed a new calculus course with short-term projects. We use Maple. We had laboratory reports for projects as well as traditional kinds of instruction, and we received NSF grants to reform those courses as well.

♦ We have many credit hours in experimental courses. Our philosophy has been to emphasize concepts, introduce technology, and do a lot of modeling. The applications in calculus and differential equations are done through projects that take a few weeks to do; they are open-ended. We use Maple throughout the campus and we are getting the support of engineering and physics. We are hoping that it will begin to take off through the curriculum. The professors like it. In differential equations we have large sections and use peer learning assistants.

Remediation

♦ We have problems with remedial classes. Fifty percent of our students take a series of computer-generated exams through the semester. All students are required to attend classes, with a maximum of 20 students per class. We provide a room full of tutors at all times (this is due to a very talented director). They just published a book with all the information an undergraduate student should know about mathematics.

♦ We offer no remedial courses. We have a problem with engineering calculus students who are ill prepared, so we have a slow-paced calculus course, which is both popular and successful. There has been a lot of pressure on the retention issue and on the issue of students failing calculus. Obtaining data on these issues is important to people like me. We are interested in whether we should go to smaller-size calculus sections. We have a very small number of math majors.

♦ In our state we talked to the regents and the department of education to let them know that students in those kinds of courses (eighth- and ninth-grade algebra) were not going to get college-level credit. This filtered down to the high schools.

♦ Placement examinations are used for all freshmen. A Treisman model program was created to address some of these problems. It has been very successful. Random interviews were held to assess success, and these show that if students put effort into learning, then they succeed. There need to be such efforts in order to get more people into science, especially for minority students.

♦ We are a private school, structured differently. We have a junior college on campus, and students who need remedial help go there. When they come out they tend to be ready for calculus. There are a lot of students who are weak in math, and we are faced with them if they are strong in English and the humanities. As of September there will be an increased requirement for two math courses (which can be any math or computer science courses above algebra and trig). This is going to create a large influx into the program, and we are trying to cope. In this institution, with the tuition so high, they actually count tuition as real money. As it is, I have approximately thirty classes a year taught by part-time people. After reading "You Are the Professor. What Next?", we got together to discuss teaching issues, technology in the classroom, and videotaping classes.

Majors

♦ The number of math majors is down. It would be nice to reverse this trend. It should be natural to choose mathematics as a liberal arts major. There are obstacles. The beginning courses require five instead of three contact hours.

♦ Retention of math majors is also a big problem. Part of it is the anonymity of the large department: we don't have a sense of community.

♦ The department is looking at the undergraduate major. The number (50–60 graduates per year) of math majors has not declined, but GPA's have. There is an undergraduate differential geometry course. Students take a two-year sequence in a core subject like topology, algebra, or analysis. This two-year sequence has had some unexpected consequences. It may explain the lower GPA's. The department is waiting to see. There is some data on high school performance versus college performance in mathematics.

♦ The numbers of math majors has increased, and we also allow people to take math majors from other departments.

♦ There are 130 majors, about half of whom are teachers. A database is kept on these students. There is an undergraduate room. The department teaches about 3,500 students per semester in freshman courses. The university has recently moved from a quarter to a semester system.

♦ We are mainly an undergraduate school. All students do two projects. The junior project is an interactive qualifying project; the senior is more major specific. In math we instituted an industrial project that students could do for the senior-year project. We developed teams, and there are three or four students working on these projects, with companies willing to pay for the work. Three companies gave us projects, and with this we are able to give release time to the faculty members because we have a deliverable project to a company. Some of these projects have turned into master's degree projects.

♦ We have the fear that our higher administration is not geared toward scholarship. We started a senior research experience that we hope will become a requirement. You could do an industrial side, or you could participate with a professor in a project. A number of students are getting together to work on something. Most undergraduate math majors do not know what research is all about.

♦ We have a very successful mathematics program: 8 percent of our undergraduates are mathematics majors. We have an REU program and bring in about 24 students each summer (one fourth from outside). The local ones are very successful in doing research and in getting research papers published. We have a tremendous number of activities all the time: math dinners, two undergraduate talks a week, ice cream socials. All senior math majors have to give talks as part of their senior year in order to graduate. We don't have a problem with calculus. Our calculus courses are the reason why we have so many majors, primarily because we have teachers who are extremely enthusiastic.

Teacher Education

♦ The department is concerned about its teacher education program in mathematics. Teachers are being produced who do not know and do not like mathematics.

♦ We are trying to do teacher preparation. We have 450 math majors overall, but the largest contingent of those are preservice high school math teachers. We teach two classes in the department to all preservice elementary teachers. All these students are seeing computing in their math courses. The elementary teachers are seeing it as an essential part of their exposure to math. We have a program that affects a small number of prospective teachers. They teach a couple of algebra and trigonometry courses for money, and they are required to take a course about their experiences in the classroom. This is an opportunity to work under the guidance of a mathematician and to think about the meaning of the math they are teaching to their fellow students. We insist that they take about one half classroom load outside of the program.

♦ We have been involved in several attempts at this kind of outreach, investigating the role of university faculty in mathematics education. With some private funds we are sponsoring three university faculty who are spending time at a local school. They set up and made available an Internet program for the students and teachers who are in the workshop.

Graduate Students

♦ We are under pressure to downsize the whole graduate program, and I will have to make the case for maintaining it. I don't really have good information about retention of graduate students. Most are there to get a Ph.D., but we often have students for six years before they leave (without a degree). We don't have exit interviews; they don't tell us they are leaving, let alone why they leave.

♦ Some data about graduate programs around the country would be very helpful. We have only about ninety Ph.D. students, and there is an amazing variation in the average time to completion, the workloads, sources of support, ultimate career goals, etc.

♦ After a year of lobbying we are starting our industrial master's degree. We made initial contact with over one hundred firms, mostly with our alumni, and they are anxious to have our students go into the program.

♦ We are currently reexamining our graduate degree, and we are trying to strengthen our master's degree, with an eye towards employment outside academia. The state cancelled a number of our courses because the enrollment was too low. We graduate 25 students each year.

♦ We are examining our master's program. While we are traditionally a Ph.D. department with very few master's students, there is a need to develop a strong master's program in conjunction with other departments.

♦ We developed a Ph.D. program with a substantial industrial component. By making connections to these companies, all of our students have some industrial component to their education.

♦ The university has decreed that first-year graduate students will no longer teach. They will be involved in learning how to teach. The central administration gave us the money to do this, which shows a commitment on the part of administration to improve the climate of the department.

♦ There is a very good sense of community between the faculty and graduate students. First-year graduate students do not teach, but second-year students participate in lecture/lab courses. The best senior graduate students run their own courses just like the faculty. There are some very good minority students.

♦ We have a very good TA training program. In their first year they only do grading. In the spring they work with a mentor in first-year calculus; they have conferences and watch. Then they are assigned one lecture, are criticized, and taped.

♦ We have had success bringing in graduate students the summer before and the summer after their first year to give them a resource seminar and some knowledge of undergraduate mathematics (so they can go into a higher-level of algebra). This has cut off almost a year from the time that they are spending in our program.

♦ We have one person who puts a fair amount of attention into TA training and follows up. We have graduate students teaching a lot; they are actually responsible for their own sections, including some experimental ones. They are getting teacher training. Also, we now have a small grant that is intended to support some graduate students going to small colleges nearby to work. We view this student teaching as a component of their education, and we view it as a continuing part of their education and training.

♦ We are having trouble finding ways to teach communication. If we don't have good vehicles for providing these kinds of skills to people in Ph.D. programs, then we don't have the competence to run the program (it is like saying, "go listen to a good opera singer, then go home and sing"). The math community needs to recognize that it should seek outside help to provide these skills. We need people with experience in this area to share with other people. We discovered that there was a committee looking at the same issue, but not for mathematics. Sometimes campus-wide teaching and learning centers that provide technical skills and push for the crucial goal of getting qualified professional people and leading faculty involved to demonstrate that this is really a worthwhile activity—that engages the senior faculty as well.

♦ We have graduate students on food stamps; they don't get free tuition. The funding at the university is year by year. We cannot offer a lot of our assistantships until March or April.

♦ We have made an effort to recruit and work with American graduate students. Personally I think that it is important to have American students so that future generations of students will be taught by people who have some sense of their own culture. I have worked very hard to recruit minority and Black students, because we need to have more traditional minority people with Ph.D.'s at institutions. In the first year the graduate student's only duty is answering questions at a drop-in center. In the second year they may teach a small section of precalculus or may even teach a calculus course. We don't do a good enough job in helping to train students to become better teachers.

♦ We have a fairly good record for attracting women into the program. However, a large percentage of the women leave. The graduate students themselves don't offer any explanations. We would like help to find out how to deal with this.

Deans

♦ The idea of a document that I can use for taking to a dean, coming from a national platform, is very attractive. We tend to think that we are much more active in calculus reform than anyone is in their respective fields. If this is true, we need to have the data available. This is the kind of thing that might excite them. The math community is doing more than everyone else. Mathematics is the key subject for lots of other things; the pervasiveness of mathematics is important. We ought to be able to show that how well our students do in mathematics makes a big difference in their later lives. We need to show the importance of mathematics.

♦ The dean was somewhat aggressive about the problem of retention and seemed to blame particularly the math department's calculus classes; he mentioned that the figures were quite alarming. We are obviously very interested in that issue, and we would like to know more about what is happening nationally.

♦ We are thinking about restructuring the curriculum. We need information, since my dean demands that we teach calculus more efficiently. The administration feels that calculus reform is cheaper, and everything comes down to dollars. We need information on what it takes to run a quality program. We need information on what it takes to teach calculus and why it is important to have calculus being taught by regular faculty. The temporary teachers and lecturers are people who couldn't make it in research careers in mathematics. These people are good at being able to get students through exams, but that's not what a university is all about.

♦ We need some evidence of what works. We proposed teaching calculus in smaller sections, but the dean reacted by saying that this is only what you guys say. We need evidence that changes will make a difference.

♦ Our dean takes essential control of vacated lines; we hire on a probability basis.

♦ Much of our ability to copy (other programs) depends on the vision and judgment of the administration. They must have a sense of the quality of the institution and its mission.

♦ We want some kind of norm. When you are talking to a dean, what does it mean for a faculty member to be productive? Our big word is assessment. We are assessing our graduate programs and found a couple of ideas in the David Report. We are trying to assess the job in teaching calculus, because the math department is being blamed for not doing a good enough job in teaching calculus to engineers.

♦ We have had increased resources. We did not make our case on the basis of teaching; it was made by candidates whose credentials glowed in the dark. We were very aggressive in pursuing joint appointments. These are nearly free if you are willing to talk to administrators at higher levels. We have gotten some outstanding people for almost no resources, and this has had the effect of enhancing our image throughout the university. You need a high-class computational system in order to do mathematics. In ten years' time you will not have a good mathematics departments if you don't have a good computational system. We made the case for this and got it.

♦ We need to find a way to convince administrations that the intellectual life of the department is extremely important and affects the way the life of the student happens.

♦ How do we respond if we are asked to justify the quality of the program? Why do we have high-quality faculty? For teaching? For research?

♦ People in senior central administrative positions are not people whose training is in the university. We spend a lot of time educating people whose view is from an MBA perspective who don't understand what a university is about. A lot of time in university committees is spent trying to educate the administration on the financial part that this is not a business, but a different kind of enterprise. It is becoming a real impediment.

♦ Money is usually gotten at the expense of someone else. Deans need to find where the money can come from.

♦ People spend a lot of time doing studies about foreign TA's, etc., but the responsibility that students have toward learning is never talked about.

♦ We are a small private university with approximately ten thousand undergraduates. We did some restructuring several years ago; we have a significantly reduced department, and we have been unable to meet our target cuts for restructuring. It appears now that there will be a second round of restructuring and that all graduate programs in the institution are going to be affected. Almost all advanced graduate courses are threatened. I would like to hear some arguments to use with my deans.

♦ Regarding a well-known chair: No one should miss the very important point that every time he sought resources he identified to the dean what he would do with them. This helps to get the resources.

Development

♦ We don't have a strong tradition of development in mathematics. All the advice we have gotten from the university I would call generic. We would like sample alumni newsletters from other departments.

♦ As soon as a student gets an award from a donor, that donor must get instant gratification. It's important to make sure a letter is written immediately.

♦ We have an awards banquet for people who are potential donors. We get them to interact with the students.

♦ For development you need to stake out your territory and decide who will work with you once you have established the contact.

Libraries

♦ We have our library in the science library, and even there we have to battle for every shelf. It might be worthwhile for the Task Force to get data showing why math departments need libraries.

♦ The hottest issue with our faculty is the issue of libraries. We have been going through the list of journals and advising which ones we can cut to allow us to bring in the journals people have asked for. This year we have to cut an increasing amount of money from our libraries. We don't have a handle on how to hold the line. We are beginning to believe that the profession needs to address the issue, since library budgets are increasing at a rate larger than the education price index. The librarians believe that if the profession addresses this at a higher level (such as boycotting certain journals that are high cost), this sort of acidity would cut down on the price of journals very quickly.

♦ We have heard cries for help for the last thirty years. Our library is not one of our biggest problems, since a former chair has put a lot of effort into our library. He started an endowment for the library, and faculty who teach an extra course can put some amount of money into the endowment — the amount that we say the course costs, or about $10,000. This impressed the administration enough that our library is in good shape.

Miscellaneous Advice and Commentary

♦ Many people are getting discouraged; there is not much interest in supporting mathematics. Something is needed to get administrations to feel that mathematics is worth it. I feel that many are downsizing.

♦ Many administrators have someone in the family who has had a bad experience in mathematics.

♦ One of the problems of mathematics is that mathematics is invisible in the political structure of the institution. Most people don't know a lot about what we do. We need to learn to speak with a common voice, and the math departments need to work to become more visible.

♦ Mathematics does a poor job of selling itself. Our initial courses should provide a good experience for students. We need to show that "math is a smart major" and that math majors make more than other science majors.

♦ The idea of a manual or training program for chairs is an excellent idea and something that will come to be. Very often chairs will come into the job without much experience; suddenly they are supposed to have a broad view.

♦ The major problem is communication between the math department and other departments. The provost has made an effort to engage the university-wide community, trying to have the math department communicate with the other departments. He is going to resurrect a committee that died in 1985 to help. The other departments need to understand the pressures on a math department; we need to make an effort to go out to the client departments to get information and feedback.

♦ Mathematics is really key to what is happening in the institution. As a discipline, mathematics is doing more thinking about the way it educates its students at all levels...than anyone else.

♦ Technology is an enhancement, not a replacement.

♦ A new faculty member in science will receive $400,000 in setup, while less than $10,000 is allocated for space for staff serving 7,000 lower-division students.

♦ It is a myth that the library expense will start to level off because of electronic journals, etc.

♦ We arranged for six of our women alumnae to meet with girls from middle schools and high schools nearby. We showed them things that you could do with mathematics, and then we had the women talk about their jobs. In this way the students really learned about mathematics.

♦ We should be concerned about our profession. There are no jobs now, but current math majors will not reach the market for many years. The number of math majors has been dropping. At every level, courses are more populated by graduate students from other disciplines like engineering and business. Undergraduates are not as well prepared as they were five years ago, yet more subjects, such as sociology, are requiring that their students understand more mathematics.

♦ We need more resources, more time, and more faculty — yes, all of them.

Chapter 6
Comments from Deans

It is a sobering experience to overhear a frank appraisal of your shortcomings. Right or wrong, the comments represent the way another person views you and interprets your behavior. When that person controls your resources and future, it is essential to understand what those views are before you can change them.

After conducting focus groups with many chairs of mathematics departments, the Task Force conducted three separate focus groups with deans of doctoral-granting institutions. There was no systematic attempt to cover all institutions or even to sample the various levels. Deans are busy people, and the focus groups were conducted in conjunction with other meetings in order to attract as many as possible. A few deans attended more than one focus group, but most came to just one. Most were anxious to express their views about mathematics, both gripes and compliments. A good many asked the Task Force for advice: How do I deal with my mathematics department? Why is mathematics different? What can I do to make mathematicians understand?

For almost every dean the corresponding chair had attended a previous focus group. While in many cases the chair and dean seemed to understand one another quite well, in some cases it was clear that the dean saw the department in vastly different ways. These were often departments in distress.

How do deans view mathematics? There isn't a simple answer, as the notes from these meetings show. Some sound exasperated, some expectant for change, some ecstatic and proud. But there are some themes that run through many of the discussions, and they are themes that are worth listening to because they represent the way administrators (and often colleagues in other departments) view mathematics and mathematicians. If they are wrong views, we need to change them; if they are right, we need to change.

The prevalent theme in every discussion was the insularity of mathematics. Mathematicians do not interact with other departments or with faculty outside mathematics, many deans claimed, and they viewed this as a problem both for research and for teaching. In many cases, deans contrasted mathematics with statistics, which they pointed out had connections everywhere. The deans spoke of a lack of "teaching dialogue" with other departments, but largely they seemed to view mathematics departments as excessively inward looking. It was viewed as a severe defect, and many deans who heard it voiced immediately agreed that it was their problem as well.

A second theme is slightly less focused but persistent as well. Mathematicians, the deans often claimed, show little interest in undergraduate education in general and remedial courses in particular. The lack of interest in remedial work seemed to ignore one of the fundamental missions of their institutions (at least for some), and there was only a passing acknowledgment by one or two that admissions standards played a role here.

Closely connected to this theme is the view that mathematicians who are interested in education have a second-class (or worse) status in the department. A number of deans recited cases in which they perceived departments had obstructed attempts to improve instruction by bringing in new faculty. They believed that departments were unwilling to broaden either hiring or promotion criteria to accommodate faculty who would improve the instructional program.

And many deans saw mathematicians constantly squabbling with one another, especially pure and applied. It was apparent that in some universities the deans had been forced to intervene, and in one or two cases had participated in dividing departments. Even when the deans merely looked on while departments argued, they viewed the divisions within mathematics as weaknesses that made hiring contentious and expansion of departments fruitless.

It is important to note that not all deans viewed their departments in these ways. One or two praised their departments for having cross-disciplinary programs. Several expressed pride in a first-rate instructional program in mathematics and commented about the exceptional reform efforts in recent years. A few believed that their pure and applied groups worked well together. But these themes—insularity, lack of interest in instruction, squabbling between factions—were present in every discussion.

There were other views expressed less often, and they show both animosity and affection for mathematics: The mathematics department is the most feared on campus. The engineers are not interested in the (reform) courses the mathematicians want to teach. Don't ask for small classes if we don't have the resources to provide them. The math faculty forget that their role in life is to teach undergraduates. The mathematics department seems to have a siege mentality (the "Rochester Syndrome"). The department feels underappreciated, under attack from students and professional colleges. There are many complaints from students, but this is because mathematics teaches more students.

And there were some deans who enthusiastically praised their mathematics departments. It is interesting to read their comments below with care to see how they measure success.

One point should be emphasized here. The comments below represent views of the deans, and they are not necessarily accurate views. But one has to deal with misunderstandings before dealing with the truth, and of course even some of the outrageous remarks capture some truth. The aim should be to understand why one dean commented angrily, "The president has said that he gets more complaints about the math courses than anything else," while another boasted, "I can't remember when I got a complaint about math!"

Comments from Deans

Insularity

♦ Concern: I worry about the insularity of the mathematics department, especially in its relationship with applied math and statistics.

♦ There seems to be a large disconnect between mathematics and other sciences, because there is very little interaction between mathematicians, physicists, and engineers.

♦ I had a mathematics department with a revolving door problem. It was very insular, with only a couple of connections to physics and no participation in the teaching dialogues going on throughout the campus. General education issues had passed them by. Our calculus classes were taught in classes of 350. Then the provost offered the mathematics department the opportunity to move from a floundering department to one of the best departments on campus. They turned around: hired different kinds of mathematicians, taught calculus in small classes, became involved in K–12 education. The department grew, and all mathematics courses are now taught by math faculty.

♦ The mathematics department does not interact well with the rest of the university. Our statistics program is all over the campus. Following an outside review, the university is moving to build a separate department of statistics whose principal focus will be on social rather than mathematical statistics.

♦ We have an outstanding statistics department, fully integrated into the university, with an interdisciplinary faculty. But the mathematics department is insular and continues to have a poor reputation with students and engineers. Some of the major complaints concern their teaching ability. The university has a president who is very concerned with student retention, but the general attitude of the department is that it is all right to have students fail mathematics. Our university pays the community college to do our remedial mathematics. Our solution was to hire a new applied computational mathematician as chair of the department. An applied mathematician makes sense for the university. He has initiated discussions with the engineering college to restructure the calculus classes for engineers.

♦ I have a very good department. They do a very good job and take their job seriously. They are trying to earn their way into general education and the idea that students should learn more than pure mathematics, and they are moving toward broadening the discipline. We have departments of statistics, bio-statistics, and agricultural statistics. There is not much interaction between the math and statistics departments, however.

♦ Our department has not been insular; they have always had cross-disciplinary interests within the department.

♦ The mathematics department is traditionally very strong, but recent evaluations have identified it as slipping from this position. The major criticism is that it is too insular and that it does not have a strong culture of support for teaching at the undergraduate level. Our department is very old, but our junior appointments have been strong, and they have produced significant efforts in reform of undergraduate teaching and education, with some calculus reform efforts. But there are continuing problems: continuing insularity, an overly inward-looking department, difficulty in the placement of graduate students. We are currently rethinking the Ph.D. program and asking for a rethinking of master's programs; in the latter there is a general resistance to dealing with math and its applications.

♦ Many of my mathematics faculty are past their prime and are mystified that the students don't identify with them. Their solution is that we must get a new kind of student. They don't value teaching or pedagogy.

♦ Problem: Mathematicians are not willing to assume responsibility for teaching enough courses to meet the needs of the college. It has not been illustrated that putting more resources into the department will fix this.

♦ We have open admission at our institution, and remedial mathematics teaching is a big part of our program. The problem is that the mathematics department does not see this as part of their mission. Our solution was to hire one teacher trained in math education, and we put in place a computer-aided instruction program, with graduate student assistants and students meeting with other students. The result was a 40 percent increase in the success rate of these students. But the mathematics department did not want to consider tenure for this position, and as a result the person was lost to another university.

♦ The mathematics department teaches some calculus courses in sections of 25 to 30. Campus-wide there is great concern about the quality of math teaching. The members of the mathematics department do not talk to each other, never mind to faculty in other departments. Insularity is very prevalent. The "pure" math faculty looks down on math educators as well as the applied mathematicians. Few of the pure mathematicians have grants. The tenure-track mathematicians don't want to teach anything below calculus, yet a third of the students have to take high school mathematics to begin. The mathematics department is the most frustrating department I have dealt with. The department is huge, and they feel they can outlast any dean, provost, or president.

♦ Concerns have surfaced that the very heavy load of calculus and precalculus is adversely affecting the major. The math major is getting the short end of the stick. We don't want to consume graduate resources in an attempt to keep up with a good calculus program. And we don't want to go from being a good mathematics program to being a good calculus school. We even teach middle school math. It is distressing to see how many new engineering students need precalculus, in spite of the fact that we are not admitting unqualified students.

♦ During the last decade our mathematics department has lost a great deal of cohesiveness. We are now working to build a sense of community back into the department. We need this in order to convince the administration to reduce the calculus sections from large enrollment to 35 students per section; we have not been able to put enough money into that effort. The teaching loads for faculty members with modest research efforts are 2 and 2. There is very little participation from the tenure-track faculty in teaching these lower-level courses, and we want all of them to participate in calculus instruction every year. Unfortunately, we find that the faculty as a whole are not interested in the undergraduate program, and at the same time the person that supervises the curriculum is a very good teacher and not such a good administrator. The mathematics department almost never considers the ability of the faculty person to teach calculus, and they never consider their effectiveness in the classroom because of language difficulties. We have a large number of low enrollment (4 to 12 students) in 90 sections over the course of the year. It seems that too many of these little special topics courses are being taught.

♦ Three-quarters of our mathematics department are pure mathematicians. The rest of the department consists of some specialists who teach only and are treated as fourth-class citizens. The mathematics educators are treated as third-class citizens. About 60 percent of the mathematicians are eligible to retire; they pay very little attention to anything below calculus, since they consider this beneath them. Many of our students have to retake high school algebra, however, and as a result a lot of people teach these students . . . but not the professors. There needs to be a group that cares about this. (Our chemistry department faculty does teach freshman level.) The mathematics department has a precalculus committee that looked into the situation and made recommendations, but they were not approved by the department. Our provost sees the large budget of the mathematics department and wonders why they are always asking for extra money for things like a resource center, and he thinks the department should raise their own money. The biggest problem is "how to change the culture in the mathematics community so those mathematicians who are doing things like teaching do not lose stature."

♦ We have been very positively impacted by increasing our unit requirements for entering high school students. We had only six sections of remedial beginning students. All students have to have had four years of high school math, with at least college algebra preparation and the recommendation that they do precalculus. We also had a math lab for a long time; this has become much more of a resource center. We started converting our faculty to using graphing calculators, only to find out that the faculty did not know how to use them. We had to get the faculty ready for this. The math lab is doing a lot to cure math phobia and graphing calculator phobia. We have several faculty members going in different directions on calculus reform. The calculus reform that is getting the most grants is so unpopular with students that engineering discourages their students from taking it. We have several people involved in other projects. We need advice on how you evaluate projects that seem to go in different directions.

♦ Mathematics departments need to be able to teach courses that address issues that are relevant to students who are not going to go into a mathematics or an economics major.

♦ My mathematics department consists of a large group, and they just do their thing. This is a problem. We want to get into collaborative learning and do workshops. We wanted to invest in a center for science education, and we asked the mathematics department to participate. Instead of taking advantage of this, they turned it down. They voted not to accept a position for mathematics education, claiming that this would move them in the wrong direction. What they wanted were additional senior scholars to give them a quick fix. They have been a major disappointment.

♦ We need to select fewer doctoral students and accept more that have inclinations towards a master's degree. We need more involvement in "undergraduate education": the senior faculty are not very supportive, and most of this effort is coming from the newer teachers.

♦ Problem: Our math and applied departments do not get along and cannot agree on goals.

♦ Our applied mathematics group resides within the mathematics department. The typical problem of insularity in a mathematics department therefore has been helped by the applied mathematicians because they naturally interact with other departments. One of our strengths is general education courses. This was initially opposed by the mathematics department, but they have since joined the effort (although there is still not a lot of enthusiasm with this part of their work). Most students take their mathematics component in either statistics or computer science, not mathematics. We are now facing serious financial problems, which has focused our attention on doing things more efficiently. We are presently teaching calculus to 60-student classes, and I've asked whether they can get away with teaching calculus to 120-student classes.

♦ Problem: The math and applied departments cannot agree on who gets calculus.

♦ We have a relatively young Ph.D. program—12 years. I asked the department to identify one or two areas of focus, to represent enough people to form a critical mass. We want to maintain a balance between pure and applied mathematics in the curriculum. This is a real challenge. They also have statistics to deal with, but this seems to be working well. The tension between applied and pure seems to be difficult.

♦ Our mathematics program was not well supported by the previous dean. Presently we have started joint hires with the physics department in an area that is growing rapidly. Our basic and applied groups work together very well.

♦ We have a fairly large mathematics department, and there are a lot of things I could talk about. Leadership is very important. The mathematics department has no focus, particularly when it comes to teaching. It seems that the leadership and the older faculty are more concerned with teaching than the young faculty are. The department is split between having an outstanding mathematician and having an outstanding teacher. There are too many research areas and not a lot of cohesion.

General Problems and Praise

♦ Our mathematics department is the most feared department on campus. There are not a large number of math majors. Many of our faculty teach service courses, and they are discouraged that they cannot teach anything more than basic courses. But they have to try to teach the students they have, not the ones you hope to have. We are trying to have mathematics be friendlier to the students.

♦ Concern: Our universities need to react to the issues of K–12 education.

♦ Our mathematics department developed good courses for teaching calculus, but the engineers say it takes too long to take all that calculus.

♦ My department is very good. We have received grants for improving calculus and algebra, and we received grants to do the same thing in the public school. We have a very hardworking, relatively young department. The department feels they are not appreciated. They have accomplished much at the national level, yet they are under heavy attack from students and the professional colleges.

♦ On the issue of small classes, we want the faculty in the mathematics department to be committed to teaching well. The idea of small classes seems to have support from faculty, and it has been seized upon both as a way to teach better and to generate resources. When I commented that I had taught classes of 400 in chemistry three times a day in my career, the comment was that you could do that with chemistry and not with math. We teach chemistry that way because we can't afford to teach classes of 40; the message is we don't have these kinds of resources. The message I am trying to send is that it is wonderful to be able to teach the small classes, but they must also find a way, with technology or other resources. Don't turn around and say classes of 40 are good; now give us the resources to do it.

♦ Our mathematics department is extremely well run, with faculty concentrating in two areas of research. They have also invested highly and are really committed to math education. We have a substantial outreach program: math day, scholarships, calculus reform, serious involvement in K–12 education. We also have a large number of American graduate students and a fair number of women. We have invested in a very big way in undergraduate education.

♦ The chair has negotiated some new resources to reduce the size of the math classes. There is a sense that the smaller classes are good, desirable, and justify the faculty necessary to keep this size. We believe in small classes to the extent that resources will allow it. When you evaluate the different departments, it isn't necessarily true that the mathematics department warrants the number of faculty and resources.

♦ The department works hard at their calculus sequence because they had dissatisfaction from physics and engineering. They do an excellent job of placement within the university so that students know where to enter the math sequence. They are inundated with students from business calculus, life science majors, pre-health professions, and it's becoming uncontrollable. There are many complaints about instruction, but this is mainly because they teach more students. Partly it is because foreign graduate students are teaching these courses. They have rigorous training for these graduate students and they are certified, but this does not make any difference if the instructor has an accent.

♦ The general success of the mathematics department is attributed to hiring quality researchers, more than average community involvement, and strong involvement in minority affairs issues.

♦ The mathematics department has just had an external review (the post-Rochester Syndrome), and most mathematicians really feel that they are potentially dealing with the issues raised there. It is very clear that the Rochester phenomenon was traumatic, and it affects a lot of our conversations. The external review was the least successful external review of my seven departments: it was a heavily proactive attempt to speak for the department on various "resource" issues, with very little criticism; the department head felt that this was somewhat cultural.

♦ We are looking at mathematics across the curriculum, and we are trying to merge calculus with other disciplines in order to have more relevance to the students taking the courses. There is extreme post-Rochester sensitivity by the faculty to the restructuring of calculus for the engineering program. Engineering is not about to provide the funds to accomplish this initiative. We finally came up with a solution: to have faculty from other departments do some teaching in mathematics, and the mathematics department review came down very strongly against this. We are dealing with some fundamental hot buttons in terms of the math faculty; everyone is stressed. The biology department had a very different reaction and accepted outside faculty. The mathematicians react against interdisciplinary compromise. Rochester has really influenced the math faculty's sense of unease and what they see as the future of mathematics—they have this siege mentality.

♦ We have a successful mathematics department. We have faculty interested in pedagogical issues, a number of middle-level faculty who are outstanding researchers and are deeply committed to pedagogy, and this has created a revolution in calculus. We never had a tradition of large classes (no more than 37 students), which made it easier to achieve. We also had a cap on the number of graduate courses that the mathematics department could offer. This meant that when we added faculty, we did not add more graduate faculty, which meant that additional hires went into undergraduate education: "the undergraduate initiative".

♦ We have a very strong mathematics department, especially on the applied side. We are very interested in supporting calculus reform. We have sections of 100, and add further support and TA's in reform sections. Classes are run through a workshop where the students work on problems. The staff helps groups working on problems, and there is a great deal of technology involved.

♦ Another major problem is foreign TA's whose English we tried to improve. In so doing, we found that it is not just the English they are missing: they need to correct the cultural differences as well. Mostly they address the problem by speaking loudly or more sternly and think the students will understand. We just had 50 percent of the chemistry students fail mathematics because they can do the mathematics but cannot transfer the knowledge to, and do, the applications in chemistry.

♦ The mathematics department needs to do a better job of screening English language skills for teaching assistants.

♦ The mathematics department needs to do a better job of training and mentoring teaching assistants before putting them in front of the class.

♦ People answer their email. Perhaps the time has come to consider things like the use of a virtual TA, where a TA is communicating by electronic means rather than sitting in a classroom. Are there ways of taking advantage of the fact that the students growing up today are really able to do this very well? Can you do this in math?

♦ Regarding the issue of having instructors—part time or otherwise—teach calculus and precalculus courses, some of these instructors are outstanding teachers, and you get more teaching for your dollars. In an ideal world we would like to have calculus in classes of 25, all taught by math faculty teaching three or four courses per year. This is never going to happen, so what do we do? Do we bring in a reasonable mix to get more teaching power per dollar?

♦ We have too many precalculus courses.

♦ A few high-powered mathematicians are constantly trying to reduce their teaching loads.

♦ The real problem in the department is a lack of community and of shared vision.

♦ The department chair often shields the faculty from reality, and in particular the faculty think that all they have to do is ask for money. They forget that their role in life is to teach undergraduates. The faculty don't understand that they have to interact with people at different levels. It is very important to awaken math faculties to the great opportunities that are available to them if they behave more like other faculties.

♦ The president has said that he gets more complaints about the math courses than anything else.

♦ Our mathematics department does a great job. By sheer force of personality and many people in the department who are really committed to mathematics education, it has become a wonderful department. Calculus reform, math across the curriculum — there is an incredible amount of stuff going on. They are getting tired, and I don't know what to do about this. Leadership has been key. I can't remember when I got a complaint about math!

Chapter 7
Comments from Liberal Arts Colleges

The Task Force held one focus group (at the Orlando Joint Meetings) for chairs of mathematics departments at colleges and universities that do not offer a Ph.D. degree in mathematics. About a dozen chairs attended the focus group. Most were from small, high-quality liberal arts colleges, but the group also included a few who had a master's program in mathematics and one who represented a two-year college.

There was also a focus group with Project NExT fellows, attempting to gain the perspective of young faculty who had recently completed their Ph.D. Most of these mathematicians are now employed in liberal arts colleges, and the Task Force specifically asked about the fit between graduate education and their current jobs. Many of their comments reinforced those of the chairs. Project NExT is a program of the Mathematical Association of America, sponsored by the Exxon Foundation. It is aimed at young mathematics faculty, helping them to build connections with the mathematics community and to develop professionally during the early stages of their careers.

The chairs reported encountering many of the same issues and problems faced by their counterparts at Ph.D. institutions. Curriculum issues in undergraduate mathematics instruction were frequently discussed, and many, but not all, were involved with some form of "calculus reform". A number discussed their efforts to incorporate technology into mathematics instruction.

Of greatest interest to the Task Force were the comments that offered insight about the differing expectations of faculty at liberal arts schools from those at a doctoral-granting department. They described what they expected from new hires, and they made consistent recommendations to doctoral programs preparing their prospective faculty. The Project NExT fellows reinforced these views, pointing out that they often received little help in teaching as graduate students.

These comments are particularly valuable to departments that are taking a close look at their graduate programs and questioning whether they could do a better job of preparing graduate students for the jobs they will most likely receive. Based on the 1996 AMS-IMS-MAA Annual Survey, for those new Ph.D.s who do find jobs, fewer than a quarter will take their first job at a Group I, II, or III department, and almost a quarter will find a job in business and industry in the U.S.

The Project NExT fellows also made recommendations about the transition from undergraduate to graduate school. There was a general feeling that universities should provide a smoother transition for students, who often are surprised at the newer (and tougher) environment in graduate school.

It is important to note that only one chair (from a two-year college) indicated that research was not required of faculty. Most of the chairs said that new faculty were expected to develop a research program, and some implied that they had very high research expectations. One chair said that faculty engaged in research mostly in the summer. From the point of view of research preparation, the chairs had no criticism of research universities, and one chair cautioned that graduate schools should not change too much.

At the same time, it was clear from listening to the chairs that teaching issues dominated the life of faculty members at their campuses and that teaching was important in making hiring decisions. Clearly these chairs were concerned that most research departments came up short both in how they prepared graduate students to become effective teachers and in how they communicated a candidate's teaching potential in a letter of recommendation.

A recurrent theme in the Project NExT comments was the need to balance research and teaching. This was something that appeared to be universal for young faculty, and while many believed they learned some lessons in graduate school, others pointed out that finding a balance was already a major obstacle for them in their careers.

The idea that successful applicants to liberal arts colleges must be prepared for a wide spectrum of teaching duties came up repeatedly. Many chairs made specific references to the importance of interdisciplinary work, often saying that faculty needed to be able to "team-teach" with a faculty member from another department or to occasionally teach seminars outside of their own discipline. It was understood by everyone that faculty at small colleges must teach a much wider variety of mathematics classes than faculty at large research institutions.

In considering job candidates these chairs looked for evidence that applicants had taught courses with complete responsibility, not just as a teaching assistant. Others looked for evidence that the applicant had experience with something other than the lecture method of teaching. The ability to articulate research to a nonspecialist and the ability to engage undergraduate students in interdisciplinary projects were cited as important. The chairs were virtually unanimous in saying that an applicant needed a thoughtful discussion of teaching issues as a part of their application if they were to be considered seriously at a liberal arts college. Chairs from liberal arts colleges stressed that at their schools the entire college is the community, not just the department. Faculty from other departments often serve on search committees, and successful applicants must be perceived as potentially good colleagues, able to get along with people in other departments, and not just good mathematicians. If there is a final point to be made about applying for jobs at liberal arts colleges or at universities that do not focus on research and graduate education, it is that successful applicants must show enthusiasm for the type of institution to which they are applying. Applicants who leave the impression that they consider a job at a liberal arts college as a consolation prize have little chance of a job offer. Both the applicant and the faculty who write letters of recommendation can do more to help their cause by making sure that the application is responsive to the school's advertisement and that applicants understand the institution to which they have applied.

Comments from Chairs of Liberal Arts Colleges

Life in a Liberal Arts Department

♦ At our school and at other four-year colleges, the focus is not on the math department but rather on the college. Tenure anxiety is high. There is significant student input for tenure and promotion cases. For final promotion, candidates need to have served the college. Colleges do not usually expect a lot of funding from NSF; research, as well as attendance at workshops and meetings, is supported by the college.

♦ Faculty are involved in a weekly teaching seminar in which teaching issues can be discussed. Faculty may teach courses other than mathematics and will certainly teach service courses. Faculty are expected to do some research, mostly in the summer.

♦ A significant number of math majors go on to graduate school, but not necessarily in math; fields like economics are popular. Other students are moving toward a career in teaching in schools. For undergraduates to be attracted to graduate school in mathematics, they need to be convinced that there are job opportunities. Undergraduate research is an expanding area.

♦ Calculus reform got a slow start at our school. There is a growth in the use of technology in the classroom. Students need to learn to read, speak, and discuss mathematics; we require students to learn to read the text. We use small groups both in and out of the classroom to help develop these skills.

♦ A high degree of computer literacy is required of our students; all math classes use computers. The discrete math class has a separate lab class; we use this to get students excited about mathematics.

♦ In our department in a two-year college research is not required. The Ph.D. is good for a salary upgrade, but not a mathless math. ed. degree. Two-year college faculty can use distance learning for their advanced degrees. In a master's degree for two-year college teaching, you need statistics, algebra, geometry, analysis, and some work outside the math department for applications material. Two-year colleges need more people who can teach in more than one discipline. In accreditation for interdisciplinary work, "math" needs to be labeled as such in order to show up correctly.

♦ In tenure decisions, good teaching is a prerequisite.

♦ Interdisciplinary courses are very important at our university. We are looking for faculty who can engage students in interdisciplinary projects and who are willing to use computers in their classes.

♦ Our faculty are expected to stay active in research. Interviewees give a talk to researchers and students. There are research seminars each Friday, with undergraduates coming one week out of four; this helps the faculty stay active.

♦ Our department was into calculus reform early, but it is still not completely integrated into the program. The program beyond calculus is traditional, and the faculty have much individual control over courses at that level. However, undergraduate research is very important, and there is a project, either group or individual, for every undergraduate. About half of our math majors go on to graduate school, but not necessarily in math.

♦ Research universities could do more in providing research opportunities for faculty at nonresearch institutions. Summer workshops or institutes or opportunities for sabbaticals in which faculty could teach some classes but also participate in the research life of the department would be helpful.

Hiring Faculty in Liberal Arts Colleges

♦ Applicants to our department need to show some enthusiasm for the type of institution it is. There needs to be evidence of involvement in teaching.

♦ Too many students coming from graduate school think of jobs at liberal arts colleges as consolation prizes. Changing the attitudes at research universities would be helpful. In addition, if faculty at research institutions modeled taking teaching seriously, encouraging students to come to office hours, etc., new Ph.D.'s would find the transition easier.

♦ Applications for jobs are read carefully for a discussion of teaching issues, since teaching is paramount in our department. Letters of recommendation should address teaching, and the candidate should have a thoughtful statement about teaching. We look for experience in something other than the lecture method. Calculus reform requires more of instructors; we ask "How much time do you think you will spend teaching calculus?" Tenure is an all-college decision, so collegiality is an important aspect of the job; candidates should show some interests outside mathematics.

♦ In hiring we look for applicants with independence in their teaching, for example, having taught a class as a TA in which they controlled all aspects of the course. There are significant research expectations of our faculty. Along with teaching five courses per year, faculty will be expected to make research connections outside the college. We have a significant tradition of faculty governance, so it is important to get faculty with interests that transcend their own field. Since there are no graduate students, faculty handle all aspects of the courses themselves, and the fact that we have honors students makes it important that faculty stay active while stepping into all aspects of a career at once.

♦ In looking at a job applicant, colleges look for energy, initiative, and excitement. Some schools ask interviewees to teach a section from the calculus book as part of the process.

♦ Letters of recommendation for applicants are frequently so dissertation oriented that it is impossible to judge the quality of the applicant's teaching and whether they could handle the spectrum of teaching responsibilities. We want letters that paint a picture of individuals: what are they like in and out of the classroom, how do they interact with students, how are they as a colleague?

♦ More attention needs to be paid to teaching at research universities; it appears that the pressure for specialization and research has intensified. Many postdocs are saying that they want more balance between research and teaching. Our department seeks faculty with a broader view and the ability to communicate with colleagues outside of mathematics. Applicants are asked specific questions about why they want to come to a liberal arts college and are asked to articulate their research to nonspecialists. The hiring committee has two members from outside the mathematics department. They look for research with undergraduates, involvement with DUE grants, especially as a PI, or the ability to write expository mathematics, for example, for the *Monthly*. One positive note: there is a good supply of strong applicants coming out now. Applicants need to be responsive to the advertisement and show an understanding of our department. There needs to be evidence that research will continue, a cognizance of teaching excellence, an awareness of improvement in their own teaching, and perhaps involvement in teaching in other departments.

♦ Graduate schools shouldn't change too much. We look carefully at teaching statements of candidates. Research is important for tenure, but less important than teaching. We look for versatility on the part of job candidates, a willingness to learn after getting the job, an ability to get along with people in other departments. We need real people with a realistic view of themselves.

Preparing Graduate Students

♦ Graduate students at research universities are too focused when they leave graduate school and therefore don't fit in well in a situation where breadth is highly valued. Interdisciplinary programs in graduate school would be helpful. I came from a program where a minor outside of mathematics could substitute for one of the qualifying exams.

♦ Research universities need a gentler introduction to graduate school for graduates of four-year colleges.

♦ Graduate schools are doing okay in research preparation, but the problem is how to keep up with the field when they have a job. Graduate schools need to do a better job of preparing students to articulate their research and to move from research to teaching. The mathematics community needs to foster the idea that it is okay to teach in a liberal arts college.

Comments from Project NExT Fellows

These were oral responses to a series of five questions posed to the participants by letter in advance of the focus group. The questions were:

1. Did your graduate school experience adequately prepare you for the teaching aspects of your profession?
2. Did your graduate school experience adequately prepare you for the research aspects of your profession?
3. How could your graduate study have been different to make your answers to questions 1 and 2 (even) more positive?
4. Do you feel it takes too long to get a Ph.D.? If so, can you suggest changes to shorten the time to degree?
5. Are there changes that could be made to make the transition from undergraduate school to graduate school easier?

♦ During graduate school, I had a 6-hour teaching load and spent eight years doing graduate work. Technology was very much a part of the courses, and I had access to technology with computerized calculus. The department is good about asking the graduate students what they are interested in and letting them teach it. Because you teach so much it might take you much longer to get through graduate school, but you are very well prepared to be on the job market. To make the transition easier, the university has now decreased the teaching load for the first year of graduate school. We also had tremendous interaction with the tenure-track faculty.

♦ I had a very good experience in graduate school. One program in particular prepared me for teaching; it was subject specific, met once a week, and each week a team of two or three presented a lecture. It was hard to listen to the criticism, but we had professors tell us how to improve our presentations and material. Our teaching load was one course per semester with some supervision. We did not have much instruction on calculus reform nor on the technology involved. It would have been helpful if we had been exposed to the different trends and encouraged to be more involved with the math community. Preparation for research depended mostly on the advisor that you had. It took me only four years to get a Ph.D., and I credit my advisor with this. It would be good to have a math student orientation.

♦ We watched a professor for the first quarter and then taught; a lot of students had ten hours to teach. Everyone gets a fixed number of dollars per class per month. There was no formal training for teachers — no reform effort — no one had heard about calculus reform. The teaching was comfortable, though. Research was done as a joint effort. We had seminars, which really helped to give me plenty of research experience. I would like to see the graduate students encouraged to attend conferences; I found meetings to be really helpful to the teaching experience. It took me seven years to finish the Ph.D.

♦ I was always teaching, from the first semester on. I got a lot of experience. They have a variety of teaching reform efforts, including Treisman-style workshops, the Harvard material, and the use of Mathematica. We only taught between 8 and 9 hours a year. I was very intimidated by seminars, because I did not think I knew anything. I was a mediocre student and it was easy to get lost, and I did not take advantage of the opportunities that were there. There was a lack of sufficient orientation: literally they said, "Don't worry too much about your teaching; worry about your mathematics." Initial advice to students is crucial, and peer advice is essential. It took me six years to finish.

♦ I started teaching during the first semester and then two hours every week for the duration. Most students grade papers during the first year. In the second year you start recitations. By the third year you can do some teaching, but it is mostly recitations. Most students have their own class in the last year. Preparation for research depends mainly on the student and the advisor. We had seminars every week, and we had 5 or 6 graduate students in them every semester. If there was no outside speaker, we had to give the seminars ourselves. The department gave us support to go to meetings, and if we gave talks, they would pay us. Length of time for the Ph.D.? Four years is good enough, and five years is too long. It takes a lot of people longer because of the comprehensive exams.

♦ I needed more teaching experience, although I learned a lot about teaching from being very involved in support. The system worked well to find an advisor, and that helped prepare me for research. But there weren't regular, frequent social events that induced the students and faculty to mingle, and that hurt. The university started to treat the students as a drain on the system. My graduate training did not prepare me for the job I now have. I continue to do research in the summers (since I don't teach then). My department supports travel to conferences and places emphasis on obtaining grants, but they are happy to have me publish a paper every two years or so.

♦ My graduate training prepared me well for teaching. I was in full control of the courses, teaching a 6-hour load. I taught a wide variety of courses; the chair for undergraduate teaching made sure of this. I found the training good at helping me to balance my time and to manage my course load. The training also prepared me for research, largely because of the exposure to people in my research field. Most of my time now, however, is spent teaching and doing service. In undergraduate school I had been part of a group that had a spotless record; we all got A's without much trouble. But that group did not measure my ability to prove theorems, which is really what is needed when you go to graduate school. It is important for departments to consider designing the transition to a graduate school program more carefully, with broader, more sophisticated course work in preparation for graduate school.

♦ Undergraduate teaching did not take place in my graduate school. But I taught 6 hours each semester from my second semester on. I taught a traditional lecture class with a pretty high failure rate, which was expected. We had no teacher supervision, and there was little collaboration among the graduate students. The U.S. students were a source of cheap labor for the university. No one monitored our progress. In graduate school I got the sense that expository writing was not for young people. This turned out to be a great deficiency in my training; writing is essential to one's career.

♦ I had no training or supervision in graduate school. The seminars were very good, although most of the time I did not understand the content. What was missing most was combining of the three requirements: teaching, research, and service. We were not prepared for service. There should be some sharing of service responsibilities with graduate students, even if it is just to share the feeling that you really need to divide your time. Presently I am being told to concentrate on my research, but I am placed on so many committees that there is no time for research. I have a feeling that there is a severe lack of structure in graduate school; I needed more milestones. I did not know when I was finished. Finally, the transition to graduate school was very difficult for me. I was not prepared for it, even though I was an A+ student. I didn't know how to do a proof. In this regard, undergraduate research projects are very important.

♦ I was totally unprepared for graduate school, but because I had done a master's degree and had been teaching for some time, the experience was not as traumatic as it might have been. It took me two years to get a master's and nine years total to get the Ph.D. I embarked on the Ph.D. program in order to get tenure, but I did not want to do research. Now, however, I love it.

♦ I gave up an industry job because I knew what I wanted to do: I wanted to teach at a small liberal arts college where they expected the faculty to be scholarly and expected the students to be good. In that sense, I am very well trained for the job. It is necessary in graduate school to teach some upper-level courses and to be on a book committee. And teaching 6-8 hours each semester taught me to balance teaching and research. My research was helped by graduate student colloquiums. In my graduate school, time to completion of the Ph.D. shortened considerably when they changed the exams and began enforcing the time limits in the graduate school contract. I believe graduate schools should change their admissions policy and allow people who have been out of school for a while to enter graduate school, since they are good risks; they know where they are going. There is a need to tighten up on the time it takes to get through graduate school.

♦ I taught most of the semesters while in graduate school. There was no supervision, no introduction to teaching, no help from the faculty at all. I got help only when I asked. My research experience was fairly good. I was in a very active area in the department, and there was a seminar every week. I talked to people about research often, and I always had someone to work with and to talk to. I don't have anyone to talk to now, and it is very hard. I need to balance teaching and research in this new environment. It would be nice if there was an orientation to graduate school, giving you a chance to talk to several different professors. It took me six and a half years to finish my Ph.D.

Part III

Examples

Chapter 8
University of Michigan[1]

The task force chose to visit the University of Michigan as a result of presentations at various focus groups by current chair Al Taylor and former chair Don Lewis. From these discussions it had become apparent that something important was happening at Michigan: the department's leadership had succeeded in making a number of significant changes in freshman instruction while at the same time enhancing and strengthening scholarly activities and graduate education.

	1995–1997 Average /yr
Students	
Full-Time Undergraduates	22,019
Junior/Senior Majors	146
Master's Degrees Awarded	17
Ph.D. Degrees Awarded	16
Full-Time Graduate Students	117
First-Year Graduate Students	22
Fall Term Course Enrollments	
Below Calculus	719 (11%)
First-Year Calculus	3,025 (47%)
Other Undergraduate	2,475 (38%)
All Undergraduate Courses	6,219 (96%)
All Graduate Courses	252 (4%)
Teaching Faculty	
Full-Time Tenured or T-track	57
Full-Time Non-tenure-track	32
Part-Time	1

We were not disappointed by our visit. It was evident that the department culture had, for the most part, changed and that administrators had provided significant additional resources to support the department's scholarly and instructional activities. In addition, there was real evidence of systemic change in the department's culture. To cite just two favorable portents: the department awarded an endowed chair to the leader of the calculus reform initiative and tenured the director of its Mathematics Learning Laboratory.

The site-visit took place on September 19 and 20, 1996. Members of the team were Carl Cowen, Ray Johnson, Barbara Keyfitz, Mort Lowengrub, and

[1] Number of full-time undergraduates in the table is taken from the National Center for Education Statistics, U.S. Department of Education, Fall Enrollment, 1996. The remaining data is from the AMS-IMS-MAA Annual Survey, 1995, 1996, and 1997. The table reports the average of all available data provided by the department during the three-year period.

Raquel Storti from the AMS staff. The department had recently moved into newly renovated quarters, which certainly gave the site-visitors an immediate positive view of the administration's attitude toward the department. This impression was borne out by conversations with administrators over the next two days.

Overall we found a culture in the mathematics department that encourages and rewards innovation, one that is well rounded, that strikes a balance between teaching and research, and that supports the work of students and colleagues at all levels. This department is deeply committed to all aspects of its mission: teaching, learning, training, and research. The most respected faculty members strongly support this holistic philosophy, and their support has made a significant difference in the attitudes of both students and faculty toward the department's responsibilities. Everyone we spoke with was committed to providing a first-rate educational experience for students at all levels. In addition, the department has established a very productive environment for its postdocs.

The largest share of credit for the changes we noted was given to Don Lewis, who served as chair for ten years. Lewis had a true vision for the department and understood how to harmonize this vision with the goals of the dean of the College of Literature, Science, and the Arts and with the mission of the university. Lewis had the twin goals of returning the department to a place among the top five research departments in the United States, a place it held from the 1930s through the 1950s, and of having the faculty take as much pride and care in their teaching as they did in their research and direction of doctoral theses. The goal to be among the top five was very attractive to faculty, and Lewis was able to channel their energy and enthusiasm toward improving teaching as well, since it was on the basis of teaching that the department would get the dean's support for its other goals. Thanks to Don's advocacy, the dean and other administrators came to think of the department as both a research institute and a teaching faculty, and they came to understand that to achieve its goals the department would need funding and support for both functions. This view was borne out in our meeting with the dean of the College, Edie Goldenberg, who took great pride in the mathematics department's achievements.

The remainder of this report is divided into several sections: the Freshman Program the Mathematics Laboratory, the Postdoc Program, the View from the Dean, and, an addendum written by Al Taylor, the present chair of the Michigan department, which provides an accounting of the incremental cost of change. It should be noted that we do not attempt to highlight all aspects of the undergraduate program. We have singled out those elements that are unusual and have helped maintain and enhance overall excellence.

The Freshman Program

At Michigan the freshman program consists of a precalculus class titled Data Functions and Graphs; the reformed calculus program; a calculus with Maple class; honors courses for science and engineering students; a course titled Calculus and Combinations, with a second semester entitled Calculus and Dynamical Systems; and a theoretically based honors course for mathematics majors.

Reformed calculus was not the only experimental program developed by the faculty. Don Lewis, as chair, was always willing to let senior faculty conduct experimental courses as long as they committed themselves to three-year involvement and were prepared to do assessments of their efforts. The faculty who developed these courses (we spoke with most of them) expressed real pride in the success of their students. Indeed, it was clear to us that faculty interest in these courses has led to a much better overall freshman program. Phil Hanlon's course titled Calculus and Combinations, followed by Calculus and Dynamical Systems, is an example of such experimentation; Lewis has referred to this class as

Math Department Lounge

"mathematics as an experimental science." Other courses for freshmen developed by faculty included Hanlon's Geometry and the Imagination; Montgomery's Problems in Number Theory; Wasserman's Maple-based calculus class; Krasny and colleagues' honors class for engineers; and the largest experiment, Mort Brown's Calculus based on the Harvard Consortium material. The College administration, impressed with these efforts, provided supplemental summer funding for course development. The chair's philosophy of giving faculty freedom and encouragement to experiment contributed substantially to the positive change in the department's attitude toward undergraduate, particularly freshman, instruction.

Most of the teaching in both freshman and sophomore calculus is done by Graduate Student Instructors (GSIs) and postdoc term assistant professors (TAPs); out of a total of 260 freshmen and sophomores, 40 are taught by tenured/tenure-track faculty. It was clear to the Site-visit Committee, however, that no matter who is doing the actual teaching in a given semester, the level of commitment by the faculty as a whole is very high. In addition, evidence of involvement and success in such teaching is taken into account in promotion and tenure decisions. The department successfully nominated Mort Brown for an endowed chair based on his remarkable achievements in reforming the calculus program.

GSIs and TAPs alike regard their experience with the teaching atmosphere at Michigan as an advantage in the tight job market, and knowledge of the Michigan Calculus is felt to be exportable. Two third-year TAPs whom we interviewed clearly were very happy with their teaching loads and with the research mentoring they had received; they felt that the three-year initial appointments were very comfortable and did not consider the term assistant professorships to be exploitative. We concluded that Michigan's method of delivering calculus instruction is ideal for a system with a large number of junior and term personnel— these young people are too mature to be satisfied with roles as "teaching assistants" in traditional calculus recitations but not ready to be instructors of record in traditional large sections of calculus or precalculus.

The special training received by all participants in the Michigan Calculus is another distinguishing feature of the program. Beverly Black, who holds a joint appointment in the university's Center for Teaching and Learning, was brought into the department to help lead this effort. In practice, all incoming graduate students, postdocs, and assistant professors participate in a week-long session before the start of the fall term. This session is well structured and is described in detail in a manual written by Beverly Black, Pat Shure, and Doug Shaw titled *The Michigan Calculus Program, Instructor Training Materials for Cooperative Learning, Homework Teams, Interactive Lecturing, Teaching Writing.* This material also provides the basis for a continual assessment program that is led by the manual's authors. See the addendum to this chapter by Al Taylor, which describes the remarkable change in professional development and assessment.

The essential idea underlying the freshman program at Michigan is to focus on learning rather than on teaching. In particular, the Michigan approach downgrades lecturing in favor of a more interactive student learning environment. Expecting that most newcomers will have received their own graduate and possibly undergraduate training in more traditional modes, the program focuses on the mechanics of delivery: organizing the classroom, identifying material suitable for lectures and for student-based discovery, and initiating student interactions with one another. This learning-centered approach may be a new one for young teachers and even for some experienced college-level mathematics teachers, but it is demonstrably one that works.

The Mathematics Laboratory

The Michigan Math Laboratory has been an important component in the department's successful approach to undergraduate learning. Its primary mission is to provide assistance to a large number of students in lower-division courses. But it also appears to have a positive effect on the math program in general, giving math majors a place to meet each other, giving them a chance to improve their skills, and perhaps making the major seem more attractive. Bob Megginson, a tenured associate professor of mathematics, has the oversight of the lab as half of his duties. (The other half of his duties includes 1 + 1 teaching and committee work.)

The lab places an emphasis on providing both high-quality tutoring and an atmosphere of efficient service. When a student seeking assistance signs in and

The Michigan Math Lab

indicates the course for which help is needed, the lab manager, who acts as a traffic controller, directs the student to a table at which a tutor handling that course has space at the moment. The tutor's job is to diagnose a student's problem, lead the student to an answer to the immediate question, and provide a new task similar to the one just conquered in order to solidify the concepts.

The tutors are undergraduates; about half are math majors. Also, each TA in a course whose students use the lab is required to spend at least one of his/her three office hours in the lab (some choose to spend all three there). Tutors are chosen on the basis of performance in math courses (up through at least linear algebra) and a short interview to determine communication skills and attitudes. Tutors are paid $8 per hour and are expected to work between 4 and 12 hours per week.

Tutors go through a four-hour training session before classes begin. The training utilizes video tapes to show good and bad examples of tutoring, and the trainees role play, with experienced tutors acting as "students". The goal of the training is to make sure the tutors get students actively engaged in solving their own problems. Tutors are expected to ask leading questions so that students work out problems for themselves rather than having the tutor dictate answers.

The lab also gives more than 10,000 "gateway exams" each year. These are basic tests of skill used to assure that students have mastered material needed for their current classes. Students can take an exam as many times as needed to pass. Students are encouraged to seek tutoring if they are having difficulty passing the exams. These exams attract students to the laboratory; once there, students realize how helpful tutors can be, and the return rate is very high.

In addition to the tutors, the math lab staff includes six student managers who work 8 to 12 hours per week and lab director Bob Megginson. The lab is open 39 hours per week. The lab enjoys adequate space, occupying a large room (approximately 3,200 sq. ft.) with tables for tutoring, individual study, and taking exams. Its funding of about $25,000 for tutors and managers comes from the mathematics department's budget.

The Postdoc Program

The mathematics department was able to convince the Michigan administration that limiting freshman calculus classes to fewer than 32 students would provide a much better educational experience. The department agreed that approximately one-third of the calculus teaching would be staffed by tenured faculty, one-third by graduate students, and one-third by postdocs (also called term assistant professors). The presence of these postdocs has been a real plus for the department's intellectual life. They contribute to "cross-pollination," and in turn they feel well integrated into all aspects of departmental activity.

The two postdocs we interviewed were in the third year of their programs, and we found them extremely positive about their experiences. The department had provided extensive training for their teaching assignments, as well as mentors for their research efforts. The goal was to help these new faculty (all holding their Ph.D.'s less than three years) adjust to a balance between their instructional and research activities.

The postdocs we interviewed saw the Michigan department as a very professional one and one that takes teaching responsibilities most seriously. The postdocs felt free to discuss pedagogical issues along with their research accomplishments; they regarded the treatment and respect they get to be no different than that accorded regular continuing faculty members.

The View from the Dean

Perhaps one of the most enlightening discussions we had during our visit was with the then dean of the College of Literature, Science, and the Arts, Edie Goldenberg. She was clearly very proud and supportive of the mathematics department's efforts and achievements. She views the mathematics faculty as one deeply committed to their scholarly endeavors but at the same time equally committed to their instructional responsibilities.

From Dean Goldenberg's perspective, leadership in the department was key to its success. Don Lewis was chair when she began as dean, and he persuaded her that Michigan could build a renowned mathematics department while making significant moves toward improvement of education, including K–12. He even brought in teams from the department to give demonstrations of instruction based on the calculus reform movement. Dean Goldenberg saw firsthand what changes were possible, and she made a commitment to helping make these changes a reality.

The changes in the mathematics department contributed to Dean Goldenberg's goals of significantly improving undergraduate education throughout the College. Her first investments were in mathematics and writing. She was delighted with the change in attitude of students and other units in the university toward learning in mathematics. The department clearly delivered on its promises, and she was more than willing to become a partner in helping the department obtain additional resources. The provost at Michigan provided several million dollars for LAS undergraduate initiatives, and the mathematics department received a substantial portion of those funds. Dean Goldenberg reiterated several times that had the mathematics department not taken its commitments and responsibilities seriously, she would not have invested in its work.

Another initiative of the dean was to link mathematics more effectively with other parts of the campus. The College and department agreed on an arrangement for joint appointments in applied mathematics with engineering and other areas; this initiative led to another expansion of the department in a direction meeting needs of many units on the campus. The changes that have occurred in the mathematics department have, according to the dean, been applauded by the School of Engineering. As a result, the engineering school has made no attempt to take over any of the mathematics courses for engineers.

The dean praised the department's undergraduate research program. Both parents and faculty colleagues around the campus have been excited about undergraduate participation in research. This program is another example of how the department has contributed positively to the College's image and reputation.

Dean Goldenberg also cited the mathematics department's leadership in assessment of teaching, thanks to a partnership with the Center for Research on Learning and Teaching. The dean has used the department's work as a model for other College units.

The Michigan mathematics department leadership has done a marvelous job of educating their dean on the role mathematics can play in teaching, research, and outreach. The time both Don Lewis and Al Taylor have taken to work with Dean Goldenberg has paid handsome dividends.

Freshman-Sophomore Mathematics—University of Michigan: An Accounting of the Incremental Cost of Change

B.A. Taylor, Chair
August, 1997

Introduction

Freshman-sophomore mathematics instruction at the University of Michigan has seen many changes in teaching and administration over the past decade. Hopefully, these changes have resulted in a significant improvement in mathematics instruction for our students. The bottom line on evaluation is, Is it better than what we were doing before? I think there is no doubt the answer is a resounding yes.

The cost of these changes was and continues to be substantial. As at any department whose faculty is concerned with quality education, change continues as we strive to improve all aspects of our educational program. Fortunately, this decade has been one in which improving the experience of freshmen students was, and continues to be, a very high priority of our administration. We have received financial support as well as encouragement in working toward our goals. The aim of this report is to record, from an administrative point of view, the nature and incremental cost of our changes to date. My best estimate of the direct costs involved is about $700,000, or approximately 15 percent of the budget expended in teaching these courses.

The largest and most expensive changes were made to the first-year mainstream calculus courses, Math 115 and 116, which have enrollments of approximately 4,300 students in an academic year. The academic and pedagogical changes for this project, which was funded in part by a grant from the National Science Foundation, have been described in a report by Morton Brown.[2]

Materials prepared for the mainstream sophomore year courses, Math 215 and 216, which enroll approximately 3,000 students each year, are also available there. A formal report on the academic and pedagogical changes implemented there has not yet been prepared.

Summary of Incremental Costs

The following table summarizes the costs of our revised instructional program. Each item is explained in more detail in the following paragraph with the

[2] Brown, Morton, "Planning and Change: The Michigan Calculus Project", in *Calculus: The Dynamics of Change,* Mathematical Association of America Notes, no. 39, A. Wayne Roberts, ed., 1996.

corresponding label. All salary costs are in terms of dollars in the 1996–97 academic year.

1. Direct incremental annual costs	
A. New junior faculty positions (9 FTE @$53,000)	$ 477,000
B. Computer labs (5@$24,000 plus systems support)	$ 180,000
C. Instructor training support	$ 30,000
D. Mathematics Tutoring Center	$ 15,000
Total	***$ 702,000***
2. Indirect costs	
A. Space charges for additional faculty	
B. Systems support for computer labs	
C. Increased workload on departmental staff and administration	
Unable to accurately estimate these costs.	
3. Startup and other one-time costs, supported by NSF grants over a six-year period	
A. Faculty release time for planning and curriculum development.	
B. Development of instructor training program.	
C. Release time for curriculum development.	
Total (grant support and matching University funds)	***$1,083,000***

To put these costs in context, I estimate that the total departmental budget attributable to the freshman-sophomore instructional program, neglecting such overhead costs as space and utilities, is about $4,849,000. Thus, the incremental costs are about 14% of the total. The estimated total costs are broken down as follows.

4. Costs of freshman-sophomore instruction, exclusive of space and other infrastructure costs absorbed by the College	
A. Salary cost of tenured/tenure-track faculty involvement	$ 1,632,000
B. Salary cost of other post-doctoral faculty	$ 1,477,000
C. Salary/tuition cost of graduate student instructors	$ 1,280,000
D. Mathematics tutoring center personnel	$ 88,000
E. Office/systems staff time attributed to freshman-sophomore program	$ 342,000
F. Staff support from the Center for Research in Learning and Teaching	$ 30,000
Total	***$ 4,849,000***

Explanations of the Incremental Costs by Item

1.A. New junior faculty positions (9 FTE @$53,000) $ 477,000

The major part of the ongoing costs are due to increased faculty needed to reduce class size and to increase support for instructors of freshman courses. In the early 1980s, class size in freshman calculus at Michigan was about 35 students, with some sections taught by faculty ranging up to 50 students. After having experimented with teaching in large lectures, medium lectures, etc., we became convinced that the best educational method for teaching freshman mathematics is the small-class format with a single instructor in charge—the smaller the better, although the size of the classes must be large enough to make the cost affordable. Over the course of the decade, our average class size has come down until it is now about 29, with no freshman class allowed to have more than 32 students. We also ran some pilot projects with class size as small as 24. The class size reduction has required us to teach about 24 extra sections each year, or 6 FTE's (full time equivalents). The remaining 3 FTE's have gone into educational administration of the courses, primarily increased support and training for classroom instructors. The additional faculty we have hired are primarily new Ph.D.'s with three-year appointments as assistant professors who teach two courses each term and are also expected to carry on an active research program. Their salaries in the 1996–97 academic year were $38,000, $39,000, or $40,000 depending on whether they had held a Ph.D. for one, two, or three or more years. Only mathematicians who have held the Ph.D. for less than three years are eligible for these non-tenurable appointments. (Only tenure/tenure-track appointments are made to those who have held the Ph.D. for three or more years, except for short term visitors.)

Problems associated with teaching many small sections are well known. First, it is expensive. The costs of having all such courses taught by tenured faculty are prohibitive. However, it is essential that senior faculty be intimately involved with and have control of all aspects of the course. Maintaining uniformity in material covered and quality of instruction is difficult and must be constantly monitored by senior faculty. With so many young instructors, many of them initially inexperienced, an extensive training and support program must be maintained. In some years we have had as many as 45 new faculty and graduate student instructors in our start-of-the-year professional development program. Providing this experience and training in teaching is an important part of our department's educational mission in supporting mathematics and its teaching throughout the country. Over the past five years 150 Ph.D.'s in mathematics, both postdocs and our own Ph.D. alumni, have been through our program. On several occasions I have received laudatory comments from department chairs at smaller institutions on the experience and attitude toward teaching of our alumni.

The costs of setting up and maintaining this instructor-development program are significant and, in my view, essential. It amounts to about 3 FTE's of increased faculty effort over our old program. To explain where the increased effort has gone, let me compare the work done now on our mainstream freshman calculus courses with that done previously. Before, each of the first- and second-term courses, Math 115 and 116, had a faculty member and a graduate student assistant overseeing the course in each term. This amounted to 2 FTE's of effort in each academic year. Their responsibilities consisted of a formidable list of tasks:

(i) Preparing syllabi and texts, preparing and oversight of the administration of uniform exams, other day-to-day administration of the course

(ii) Advising instructors on best practices in teaching, holding meetings for instructors to coordinate sections

(iii) Monitoring the quality of classroom instruction, working with individual instructors on methods of improving instruction

(iv) Monitoring the effectiveness of the course for students

(v) Working with client departments to make sure the syllabus is appropriate for their students. Consulting with them on curricular issues and changes

(vi) Monitoring national curriculum development to bring improvements to the Michigan program, e.g., in integrating technology into the curriculum

(vii) Dealing with student complaint and disciplinary actions

(viii) Coordination with leaders of other freshman-sophomore courses to make sure the courses mesh properly over the two-year program

It is pretty tough to see how two faculty members, each assisted by a graduate student, can operate the mechanics of teaching over 140 sections of courses with 120 different instructors and 4,300 students, and still find time to carry out all these other tasks. Further, each faculty member was also expected to teach a course each term, keep working with graduate and other advanced students, and maintain his program of scholarly research and publication. Indeed, only a superhero could keep up with the expectations of such a job.

Recognizing this, we now have about two additional FTE's of effort that go into supporting Math 115/116, with the third incremental FTE of effort being put into the second-year program. One of these is split among experienced faculty, postdocs, and graduate students who visit classes and generally assist in instructor training on an ongoing basis. Another is split between two faculty members who have time to consult with colleagues in other departments and think about the long-term development and evaluation of our efforts. The FTE on the second-year courses goes into writing and coordinating the computer labs and oversight of the graduate student instructors who assist in the labs. In addition to this, we also have the half-time assistance of a staff member from the University's Center for Research on Learning and Teaching, who assists in instructor training, works with instructors in the classroom, and assists in evaluating the effectiveness of the program. While in the early stages of a curriculum development effort one can count on extraordinary efforts of talented individuals to make good things happen, to maintain educational improvements, one has to have a structure in place where jobs can be done and rewarded on a basis commensurate with other departmental work. Further, it has to be realized that a "half-time" assignment to such a position should not be a 20-hour-per-week job unless the faculty member is also released from the normal expectations of maintaining research and other scholarly activities. Sufficient support of the most important person in teaching, the classroom instructor, is essential if quality is to be maintained.

While the costs of our program are substantial, there are also significant benefits that accrue to the individual instructors, to the department, and to students. When I came to Michigan as a new faculty member—and, indeed, when I first taught as a graduate student in the 1960s—the support given new instructors was almost nonexistent: "Here's the book, the syllabus, and there's the classroom. Bring us your first exam so we can check it over for you. Come to see me if you have any problems." The difference today for faculty and graduate student instructors is remarkable. Before classes start there is a week-long professional development program where they are given extensive training on the goals of the course, the goals of the student population in the course, and suggested teaching methods for helping the students reach their goals. All faculty, even senior faculty, who have not taught the course recently go through this program. Throughout the term there are classroom visits and regular meetings with other instructors to discuss problems and ideas as well as coordination with other sections. Materials describing our instructor program, developed primarily by Pat Shure and Beverly Black, have been published by Wiley.[3] New teachers very rapidly come to appreciate the role of mathematics instruction and its importance in a university setting.

For the department, the incremental positions have brought us the benefit of more bright young faculty full of ideas and enthusiasm. They enrich the scholarly life of every mathematician (and colleagues from other departments with whom they interact) at Michigan. Part of our professional development program for young faculty focuses on the necessity of maintaining a balance of work in teaching and in scholarly research, both of which are essential to a successful academic career.

For the students in elementary mathematics courses, they have the advantage of supportive and enthusiastic instructors who are experts in the discipline and who have an appreciation of mathematics and its wide range of applications. The small class size allows them to get to know and work individually with their instructor. For more advanced students of mathematics, there are more experienced faculty available to talk about and work on individual questions and research problems.

1.B. Computer laboratories

We have five computer laboratories, each consisting of fifteen UNIX workstations that are used in the sophomore-year courses and for several other courses (e.g., honors and upper division). We estimate the labs have a three-year lifespan and the equipment and ancillary charges of replacement are about $60,000. There are additional costs for having student monitors in the labs, approximately $4,000 per lab per year, and we estimate the labs require at least 40 percent of the effort of our two departmental computer systems staff, or about $50,000 per year. We were fortunate to have received the machines in three of our labs as a gift from the Hewlett Packard Corporation in support of our work to integrate technology (MAPLE) as a key part of our multivariable calculus course.

1.C Instructor training support

This item is mostly the half-time effort of a staff member from the Center for Research on Learning and Teaching. There has also been some summer sup-

[3] Black, Beverly, Shure, Patricia, and Shaw, Douglas, *The Michigan Calculus Program Instructor Training Manual,* John Wiley and Sons, 1997.

port for those preparing the professional development program for new faculty and graduate students.

1.D. Mathematics tutoring center

For many years the department has had a tutoring center, called the Math Lab, for students enrolled in freshman-sophomore courses. Since the move to our new facility in East Hall and in support of our increased use of "gateway" or "mastery" exams, the number of student visits per academic year has increased to over 20,000. The increased cost of undergraduate student tutors in the lab is about $15,000 per year. We continue to operate the lab with the same amount of faculty and graduate student support.

2.A,B,C Indirect costs

Whenever the number of faculty in a department increases, there is a corresponding increase in all infrastructure costs: more offices, more secretarial support, more computers, bigger phone bills, more activity of all sorts. It is very difficult to quantify these costs, since they are not charged directly to the department. There is also a discernible increase in faculty workload. The postdocs we hire are not required to fulfill departmental service obligations. They are expected only to do excellent teaching and excellent research. So, the service obligations of a larger faculty (more than 90) are spread over the tenured/tenure track faculty (55). In addition, hiring several postdocs each year (around 12–14 in recent years) requires a large amount of faculty and administrative effort. Care is taken to see that there is a senior person working in the area of each postdoc who can serve as a mentor, so it requires senior faculty members to read applications and come forward with their recommendations.

3. Startup and other one-time costs

The courses that have been affected by the changes under discussion here involve about three-fourths of the student credit hours that we teach in each academic year. Thus, these changes have been extensive and required much work and preparation. Diagnosing and working to improve our courses was a constant goal of Morton Brown since he took on the role of associate chairman around 1980. He deserves great credit for recognizing the need for improvement and for attacking the problem long before resources were available to support changes. These resources arrived with an NSF grant along with matching funds from the College of Literature, Science, and the Arts, which allowed for faculty release time, some support for graduate students, the ability to hire consultants for evaluation and advice, and trying out smaller class sizes (such as 24 instead of 32). The College also has supported dedicated, specially furnished classrooms that support the active learning methods used in Math 115/116. Important planning and much detailed work was carried out by faculty members Morton Brown, Patricia Shure, and Robert Megginson, and by our CRLT consultant, Beverly Black. They were assisted along the way by the active cooperation of many senior faculty who devoted time to learning about the project and teaching in it. The support of the department chair, Donald J. Lewis, who pushed for the project and was persuasive in obtaining the support of the administration, was crucial.

Lewis's support was also instrumental in supporting several curriculum development projects in the department and, indeed, in creating an atmosphere within the department where such work is recognized as valuable, even essential, to meet our obligations as educators. One of these projects that had substantial startup costs in faculty release time led to the current form of our second-year courses. The idea of computer labs using MAPLE to assist in teaching multivariable calculus was started at Michigan in the late 1980s by John Harer (now at Duke University) and C.K. Cheung (now of Boston College), who have since written a book using their ideas. When they left Michigan, the program was taken over and modified by Estela Gavosto, now at the University of Kansas, and Alejandro Uribe. Our current program, which we believe is quite successful, has the form they created. Their materials will be disseminated soon, supported by an NSF grant. Some is currently available on our home page:

http://www.math.lsa.umich.edu.

Many of the startup costs and all of the large project startup costs have been supported by external grants. While departmental funds can begin pilot projects, the total costs of a full-blown effort involving evaluation and consultants needs significantly more external support than a department can supply with its own resources.

The amount estimated for this item, $1,083,000, is the total of the NSF ($500,000) and matching university funds ($583,000) that were put into the two projects: Math 115/116 (freshman-sophomore calculus), total of $963,000 and Math 215 (multivariable calculus), total of $120,000. While the department has spent more than this on curriculum development during the past decade, the remaining work has been supported by the departmental budget on an ongoing basis.

Chapter 9
Oklahoma State University[1]

We selected Oklahoma State University for a visit since it is a prototypical land-grant university, characterized by the changes and challenges of public institutions, with a reputation for a mathematics department that is highly regarded within the institution and active and visible within the research community. We wanted to understand the reasons for what, from a distance, seemed a very successful undertaking both locally, in undergraduate education, and nationally, at the research level.

	1995–1997 Average /yr)
Students	
Full-Time Undergraduates	12,985
Junior/Senior Majors	58
Master's Degrees Awarded	2
Ph.D. Degrees Awarded	3
Full-Time Graduate Students	36
First-Year Graduate Students	9
Fall Term Course Enrollments	
Below Calculus	1,638 (35%)
First-Year Calculus	964 (21%)
Other Undergraduate Courses	1,985 (43%)
All Undergraduate Courses	4,587 (99%)
All Graduate Courses	53 (1%)
Teaching Faculty	
Full-Time Tenured or T-track	32
Full-Time Non-tenure-track	7
Part-Time	5

The University and the Department

Oklahoma State University is a land-grant university, sharing and competing for the role as the major higher education institution of that state with the University of Oklahoma. Some statistics sketch a picture of this institution. For 1992–93 it reported a faculty complement of 1,114 faculty, of which 76 were part-time, and a student body of approximately 19,000; it awarded 2,710 bachelor's degrees, 713 master's, and 227 doctorates. As these figures indicate, its educational activities are broadly distributed from undergraduate to graduate in-

[1] Number of full-time undergraduates in table is taken from the National Center for Education Statistics, U.S. Department of Education, Fall Enrollment, 1996. The remaining data is from the AMS-IMS-MAA Annual Survey, 1995, 1996, and 1997. The table reports the average of all available data provided by the department during the three-year period.

struction. Typical of a land-grant university, a notable proportion of its activities are in "land-grant" areas of instruction, such as education and agriculture. The College of Arts and Sciences encompasses 37% of the faculty and 42% of the instructional load (a weighed combination of instruction at different levels) of the institution.

The Department of Mathematics, one of the units of the College of Arts and Sciences, had a faculty complement of 32 (of which 3 were part-time), approximately 45 graduate assistants funded by the state budget, and an additional group of approximately 60 undergraduate student employees. The total number of declared undergraduate majors was 78; it awarded 18 bachelor's, 10 master's, and 1 doctorate. Not unexpectedly it provided a high volume of undergraduate instruction, especially at the freshman and sophomore levels (a total of approximately 30,000 credit hours). These figures indicate a typical department within a public research land-grant university with a significant and rather high component of service teaching, a small complement of undergraduate majors, and an active graduate program. An analysis of the budget and activities of this department yielded that its instructional faculty represents approximately 3.4% of that of the University, its budget 4.6%, and its school credit production 6.5% (but when weighed by level of instruction, 4.4%). These figures suggest that the department has a relatively high teaching load but that the University administration has financially responded well to its instructional demands and to its scholarly and research activities. These tentative conclusions confirmed the reasons that motivated the site-visit: the department was successful in making an appropriate case for itself within the University. What were the reasons for this success?

At the conclusion of the visit, the visiting team felt that the department and its leadership had successfully managed, on the one hand, to respond to the mission and demands of the institution in undergraduate instruction and, on the other, to its ambitions in research. Further, it had, over an extended period of time and through stable leadership, ably communicated these successes within the administration of the University, where it was perceived as a highly collegiate unit, devoted to high scholarly and instructional standards yet entrepreneurial and aggressive in research and education. Central to this success, it appeared, were conscious efforts to leverage scholarly and educational activities on each other; to devise effective and efficient means of instruction at the undergraduate level that were highly regarded not only by students but especially by faculty in other colleges and departments; and through an explicit policy, to focus on a few areas of research.

Faculty and Research

The department sees itself as (and is) a research department. Research activities are mostly focused on a few areas in pure mathematics that are well supported by competitive grants. The graduate and postdoctoral programs are, as expected, similarly focused. The graduate program is relatively small; its further development is regarded by the faculty as one of their challenges. Its doctoral graduates predominantly enter the teaching profession at four-year institutions. As is the case in many research departments, the quality of the graduate students and their preparation is not commensurate with that of the research faculty. The graduate and research environment is enlivened by a postdoctoral program, with a cohort of four young mathematicians of excellent pedigree in the areas of departmental research focus. These postdocs are involved in the teaching functions of the department, developing their teaching skill, but having significant time for research and scholarship. The department is acknowledged to be an excellent environment for their further scholarly development.

The OSU Math Building

The departmental faculty is concentrated in two areas: mathematics education and some specific areas of pure mathematics research. The relatively small size of the faculty has prompted the department to follow a policy of concentrating its activities in a few areas rather than attempting to provide coverage of broad fields. It has also very successfully developed a very good atmosphere of mutual respect and support between those members of the faculty whose major interests are in educational activities and production of educational materials and those at the forefront of mathematics research. The end result is a high level of activity and of publication in both areas. In the education area the faculty has developed and published textbooks on the use of technology and mathematical software in conjunction with calculus instruction; they have been awarded numerous grants from the National Science Foundation for instructional activities at the calculus and precalculus levels and have engaged in consortia for major "reform" projects; they are actively involved in outreach activities to the K–12 system of the state. In this arena the faculty has leveraged its commitment to instructional activities within the University and outreach activities within the state with projects supported by peer-reviewed external grants that give it internal credibility and national visibility. At the research level the focus on scholarly activities is on a relatively narrow set of mathematical subfields in number theory, representation theory, algebraic topology, and analysis; this focus has resulted in the department developing a national (and international) reputation in these areas and considerable visibility.

At the cost of mathematical breadth, given its size, the department has developed concentrations of activity of a critical mass that have enabled it to secure significant resources from external grants in a highly competitive environment. The externally funded research and educational activities of the department brought in, in 1994, approximately $900,000 per year, more than one-third of the funds provided from state budgets. This is a notable figure that speaks, on one hand, of payoff on focus, but also on determined efforts to aggressively pursue national funding and visibility in focus areas of education and research. The scholarly activities of the department are underscored by a small, equally focused postdoctoral program that attracts young mathematicians of excellent pedigree, enlivens the research environment, and gives further national visibility to the department. This visibility is very much prized by the University administration, which is well informed about it and very supportive, as testified by the award to a member of the department of one of the few chaired professorships available to the University administration.

Instructional Programs

The research success of the department could not be sustained within an institution like Oklahoma State University without a successful program of undergraduate instruction, for the "service" component of the department provides the base on which its other activities are built. Three components of undergraduate instruction were noted by the site-visit team for particular attention: the quality of service and general education courses, the significant involvement of departmental faculty in programs of teacher education, and the role of the Mathematics Resources Learning Center.

Typical of departments of mathematics, less than one percent of the instruction in the freshman and sophomore years is devoted to its own majors, and instruction at this level represents approximately 80 percent of all registrations in mathematics courses. The department at Oklahoma State University, through the component of its faculty, where interests center on education and educational research, has developed a very good set of innovative course offerings that are well regarded by students and by the "customer" departments. Notable to the site-visit team was the fact that, in discussions with senior administrators of the University, the instructional program of the department was praised; this is seldom the case. The department has aggressively developed a set of elementary mathematics courses for general education purposes, transforming the standard remedial college algebra courses into innovative precalculus courses; it has produced textbooks and manuals, and exploited the use of calculators. Notable is the wide appeal to students of a general education course entitled Applications of Modern Mathematics, based on COMAP's textbook *For All Practical Purposes*. The significant effort the department has put into its general education courses has attracted significant external funding and simultaneously has responded positively to the needs of a large student population whose interests and abilities in mathematics are limited. This effort is particularly prized by the senior administration of the University, highly concerned with the retention of beginning students. Equally entrepreneurial and innovative have been the efforts, again led by those faculty members interested in education and educational research, with innovative and well-executed developments in the calculus and subsequent courses in differential equations. The organization of these courses is thoughtfully planned, there is a thorough program for graduate students to prepare them to teach such courses, and a significant infusion of technology is evident. Mathematica, Matlab, and Derive are integrated into these courses. The department has astutely involved the customer departments and their faculties (in engineering, the sciences, and business) in the development of these courses, resulting in a sense of ownership and satisfaction on their part. The attention and effort that the department has devoted to lower-division instruction has resulted in a very good program, external funding, and high regard within the University; it has also resulted in a level of financial support from internal budgets that would be most unlikely otherwise. That a component of the departmental faculty is strongly devoted to educational research efforts was most important to such success.

A second notable aspect of undergraduate instruction is the significant involvement of departmental faculty in instructional programs, centered in the College of Education, in the preparation of elementary and high school teachers. Many of the faculty within the department whose interest centers on mathematics education also hold appointments in the College of Education. What was striking

The Mathematics Lab

to the site-visit team, however, was that undergraduates majoring in mathematics education looked to the Department of Mathematics as their home, not the College of Education; they constituted a significant component of the departmental undergraduate student body and looked to the faculty of the department as their mentors and advisors. Note that at many public universities, secondary mathematics education majors are counted as a component of the mathematics department's majors.

A critical part of the successful undergraduate program of instruction is the Mathematics Resources Learning Center (MRLC). This Center is dedicated to the support of undergraduate and outreach educational programs. It is a large complex, capable of easily accommodating up to sixty students; it is equipped with approximately forty networked computer terminals with appropriate software, and it is staffed mostly by undergraduate upper-division students, managed by the senior staff officer of the department and recent graduates. The Center serves a number of interrelated purposes. It is a tutorial center where lower-division students receive help from more advanced students; it is a place for access to software, technology, and tapes and visuals associated with courses; it is a place for students to do homework and engage in collaborative learning; and it is an inviting place to study mathematics, with help accessible as needed. It is also the centerpiece of outreach activities to the local K–12 mathematics education community. By all measures, the Center is very successful in its tutorial and technological tasks; it clearly provides, through well-trained and managed undergraduate tutors, effective and efficient instructional support. Three aspects of the Center struck the site-visit team. The first was that many of the student tutors were students in the College of Education, not mathematics majors, yet they exuded a sense of pride and of closeness to the Department of Mathematics, clearly motivated through their involvement in the Center and in its teaching functions. One of the staff members of the Center, a recent graduate of the College of Education who planned a career in teaching high school mathematics, stated that he regarded the Center as his future point of contact with the University, and of referral for his high school students. Through the Center the department has ably appropriated as quasi-majors a number of students in the College of Education—a number larger than the number of its own majors. Secondly, the Center has leveraged this attraction of education students into an effective outreach program to high schools and into close contact with high school teachers. As a result, the department is viewed very positively in the high school community, and the Center hosts a number of high school mathematics competitions and teacher-training programs. Finally, the Center has become the center for the interactions of undergraduates registered in mathematics courses: it is the visible and accessible face of the department, and it is a welcoming, helpful, and friendly face. There is a palpable good feeling on the part of students for the Center, and great pride and loyalty on the part of the student tutors. Senior administrators of the University are fully aware of the Center and prize its contributions. The Center is, in the view of the site-visit team, an activity most worthy of emulation because of its effectiveness and efficiency in instruction, as a test bed for technological innovation and outreach activities, and as demonstrable evidence of the central role and importance of mathematics in undergraduate instruction. It is also an effective and economical means to demonstrate the commitment of the department to undergraduate students.

The mathematics major in the department is not significantly different from the standard one, but the role of the department in the preparation of K–12 teachers, as noted above, is very significant. The department has a long tradition, dating from the '60s, of strong involvement in mathematics education and of outreach activities directed to the K–12 system of the state. This tradition has been sustained through the evolution, dating from the early '80s, of the department from a mostly teaching unit into a research department. This evolution was managed adroitly, resulting in retaining a commitment to undergraduates—be they majors, students in the teacher-training programs, or majors in the sciences, engineering, and business—as the department attracted new faculty with strong interests and commitments to research and graduate education. The success of this evolution is palpable in the sense of mutual respect and support between members of the faculty who see themselves as educators and those committed to being at the forefront of research. This respect and mutual support underpin the quality and commitment to undergraduate programs.

The Mathematics Resources Learning Center is the focus of interactions in undergraduate education; the teas and coffees held in the commons of the department, especially in conjunction with seminars and colloquia, provide interactions for graduate students, postdoctorals, and faculty. The environment of these is highly collegial. The leadership of the department has successfully nurtured the growth of an atmosphere that is supportive but intellectually demanding, and a sense of community with high respect for a diversity of talents. The site-visit team could not but be impressed with the level of morale in the department.

Relationships within the University

The site-visit team early perceived that the department and its leadership had been very successful in communicating and establishing good relationships with other units of the University and with its senior administrators. This success was clearly based on real results: its commitment to undergraduate education and its visibility in research. But beyond this reality, the site-visit team noted that the department consciously and ably communicated these, placed considerable energy in preventing isolation from other units, and consistently involved faculty in other departments.

A key factor explaining the high regard of the department, in the view of the site-visitors, was the long-term leadership of two chairs who have astutely devoted considerable energy to interactions outside the department and have consciously treated other departments and senior administrators as their "customers".

Notable, for example, is the colloquium in the department in which faculty from other units are invited to speak about problems of a mathematical nature in their research. Chemists, physicists, and engineers have through this means been brought in contact with departmental faculty who, given the nature of their research in pure mathematics, are unlikely to engage in interdisciplinary research projects. The result has been, besides a few cross-departmental research activities, a very good level of interactions, a broad understanding of the department by other units, and a sincere appreciation of its contributions. This interaction has been furthered by the conscious involvement of the customer departments and their faculty in the design and evaluation of undergraduate courses. This investment by the department, and especially by its leadership, has had a significant payoff: faculty in other units speak knowledgeably about the department and of its contributions to their students and to the mission of the University.

Equally impressive was the rather detailed knowledge by senior administrators at the University of the contributions, successes, opportunities, and problems within the department. The leadership of the department had clearly done considerable work to communicate these in a realistic manner. The result was that senior administrators viewed themselves as allies of the department and were personally proud of its successes and concerned with its problems. "They are batting 1.000," a dean stated to us, then elaborated on the excellent morale and collegiality of the department and on his efforts on behalf of its graduate program, under attack by state officials concerned with its small size.

Central to the high regard of the department is that it has astutely aligned its activities to the mission of the university and of the perceived needs of other units in the institution. The ambitions of Oklahoma State University are centered on the quality of the undergraduate education it provides the citizens of the state and on its reputation and contributions as a national research university. The department has ably addressed both of these ambitions with limited resources. It has also seen to it, through astute communications and interactions, that its successes in a subject that is central to both education and research are seen as integral to those of the university.

Concluding Remarks

Oklahoma State University is not a particularly well-funded institution, especially among research universities. Yet, within limited resources, the Department of Mathematics has succeeded in developing a highly regarded undergraduate program and vital and nationally visible research activities. Long-term leadership and commitment to long-term strategies based on alignment with the local educational mission and national visibility in research played a key role in this success. Important also is the entrepreneurial and innovative nature of undergraduate educational activities that can be stimulated within a research department conscious of the importance of its educational mission as a necessary base for its scholarly ambitions. Lastly, significant efforts devoted to communication with other departments and units and with senior administrators has undoubtedly been an important element in the success of the department.

Chapter 10
University of Chicago[1]

The Task Force chose to visit the University of Chicago because several Task Force members had favorable information about the culture and esprit de corps of the mathematics graduate program that prepared future faculty to be both good researchers and good teachers. The mathematics department was known to have a very successful undergraduate program, a number of outreach programs to Chicago schools, and a new entrepreneurial Master's Program in Financial Mathematics.

	1995–1997 Average /yr
Students	
Full-Time Undergraduates	3,515
Junior/Senior Majors	58
Master's Degrees Awarded	24
Ph.D. Degrees Awarded	13
Full-Time Graduate Students	91
First-Year Graduate Students	14
Fall Term Course Enrollments	
Below Calculus	43 (3%)
First-Year Calculus	895 (58%)
Other Undergraduate Courses	458 (30%)
All Undergraduate Courses	1,395 (91%)
All Graduate Courses	139 (9%)
Teaching Faculty	
Full-Time Tenured or T-track	33
Full-Time Non-tenure-track	15
Part-Time	1

The site-visit took place on October 21 and 22, 1996. Members of the team were Carl Cowen, Mort Lowengrub, Alan Newell, David Vogan, and Raquel Storti from the AMS staff. Our mission was to explore those activities which stood out from the norm, programs we might all learn from and in some cases emulate. The reader should understand that we were not there as general critics. Like all departments, the University of Chicago has its weaknesses and idiosyncrasies.

[1]Number of full-time undergraduates in table is taken from the National Center for Education Statistics, U.S. Department of Education, Fall Enrollment, 1996. The remaining data is from the AMS-IMS-MAA Annual Survey, 1995, 1996, 1997. The table reports the average of all available data provided by the department during the three-year period.

The Department and the University

The department, consistently ranked among the top five in the nation, has a distinctive structure. There are presently twenty-eight senior faculty members, almost all of whom occupy professorial rank with one associate professor. They have eight assistant professors, one research associate, thirteen Dickson Instructors, and one senior lecturer. The prestige of the department has positive consequences for junior faculty, who tend to be highly sought after for permanent positions elsewhere. The many long-term visitors enrich the environment in visibly concrete ways. Although the department has an applied component (it offers, through the computational and applied mathematics program (CAMP), interdisciplinary tracks leading to the M.Sc. and Ph.D. degrees), its culture is strongly oriented towards what would traditionally be called pure mathematics.

From its inception in 1893, the University has been at the forefront of graduate education in the United States. The current total enrollment reflects the commitment to graduate education. Almost half of the approximately 12,000 degree students are enrolled at the graduate level. This includes the professional schools. The College of Arts and Science, which reflects that balance, is now continuing to increase its undergraduate enrollment, presently 3,500. This means an increased obligation for the Department of Mathematics, which is the single largest provider of instruction. The instruction is carried out by both faculty members and graduate students. Graduate students who teach are given lecturer rank. Perhaps uniquely among U.S. universities, only 60 percent of faculty teaching time is devoted to the undergraduate level. Undergraduate mathematics at Chicago at all levels intentionally retains a graduate student emphasis. The department determines the quality of a graduating class of undergraduate majors by their later performances at Group I graduate schools and in their academic careers. Of Chicago's graduates in mathematics, about 50 percent go on to Ph.D. programs in mathematics, and about 25 percent to Ph.D.'s in other disciplines. Much of the success of the undergraduate program is attributed to small class sizes and a well-trained cohort of graduate student teachers. Indeed, this culture was developed as far back as the early seventies when Felix Browder, the chair at that time, negotiated with the University a plan for more graduate positions in return for a cast-iron agreement concerning the training and mentoring of graduate students, about which we will talk more later. The success of the small-class format, along with the careful training of graduate lecturers/assistants, is consistent with the reasons for success we found elsewhere.

Mathematics at Chicago is taught the old-fashioned way. For the most part, there are no concessions to the movement towards the introduction of computation into the curriculum, although some compromises to this policy have been made for science courses given to physics and chemistry majors. Moreover, there is also recognition that a majority of those mathematics graduates who decide not to go on to graduate school will end up in the worlds of accounting, finance, and business, and because of this they are about to introduce an option in mathematical economics.

Candidates for the graduate program are chosen carefully and then supported generously and enthusiastically. The aim is to bring in about fifteen per year.

First-year students concentrate entirely on their studies. Second-year students begin to become involved in teaching and by the third year are fully involved in teaching about three courses per year, one course per quarter. The average student takes five years to complete the Ph.D. program. The dean strongly supports the graduate program, and there are no plans to discourage graduate enrollment unless it is demonstrated that Chicago graduates are having troubles in the job market.

The Graduate Program

The program enjoys an esprit de corps way beyond that at most universities. How is this achieved? The first answer is that the students are carefully chosen in the initial instance and then are made to feel extraordinarily special and welcome. They come with the attitude that they must work and work very hard, and the expectation is that in return the department will nurture their development as

Tea in the Math Department

mathematicians. The second answer is that all students bond with each other and with the department during a first year baptism of fire consisting of three year-long course sequences in Algebra, Analysis, and Topology and Geometry. The courses involve an enormous amount of material and homework and are fairly rigidly structured. They are usually given by nine different professors. They also encourage a spirit of genuine cooperation among students, a spirit which is initially based perhaps less on altruism and more on sheer survival instinct. During this period the students have no obligations other than to attend to their own learning. The spirit of collegial cooperation engendered in this first year seems to stay with students throughout their graduate studies and manifests itself in continued interest in each other's progress and in many student-sponsored activities such as weekly "pizza seminars". The third answer is that everybody is involved in teaching and is carefully nursed into the teaching process throughout a well-organized second year, and monitored continuously thereafter. They begin their second year by sitting in on the classes they will eventually teach, then by handling tutorial sessions and lectures on an occasional basis, and finally by taking full responsibility for their own class. Undergraduate classes are small, about 30–35 students, and are regularly visited by faculty mentors. Examinations that are set by novice lecturers must pass the eagle-eyed scrutiny of the uncompromising Paul Sally.

There are no qualifying examinations as such. Students are introduced to the research culture by taking on two separate projects during their second and third years and making oral presentations. These serve as "qualifiers" for the Ph.D. Once students pass this stage, they begin dissertation work. Students chosen for the program are expected to succeed, and through hands-on mentoring and nurturing are given every opportunity to do so. Financial support is guaranteed. The Task Force recognizes that the University of Chicago is singularly blessed by having access to the best young minds and having more-than-average financial resources for its graduate programs. Nevertheless, these ingredients alone do not guarantee success. It is the clear statement and consistent application of its own distinctive policies; a fairly rigid core structure; a nurturing, collegial and caring environment; attention to training in teaching as well as research; and the installation of a feeling of confidence in, and the expectation of, good things from every student admitted that makes a good program work. We saw a good example of such a program at the University of Chicago.

Educational and Outreach Activities

The University of Chicago Mathematics Department has a very clear commitment to excellence in undergraduate education at a level that is rare for a department rated so highly for the quality of its research faculty. Both in the teaching of undergraduate classes and in the careful mentoring of graduate students as classroom instructors, the faculty is very much involved. The program of study is unusually rigorous, and it is a point of great pride in the department that its best undergraduates are given a diet of courses that is intellectually rich well beyond what might be expected, even in an outstanding department. The department is also aware of the fact that it is the unit in the University that teaches the most, and it is dedicated to providing a high-quality education for those many students who require mathematics in their course work but are not mathematics or even physical science majors. Among the faculty, five have won the University's Quantrell Award for excellence in undergraduate teaching, the oldest prize in the nation for college teaching.

The director of undergraduate studies is Paul Sally, a University of Chicago phenomenon and a man of formidable presence and commitment. Bob Fefferman, the department chairman, introduced him to us as a kind of local miracle who combines a deep respect for the role and purpose of research with an equally strong commitment to undergraduate education and all the care and attention that the molding of a quality learning environment entails. The associate director of undergraduate studies is Diane Herrmann, who holds the (nontenured) position of senior lecturer and who plays an absolutely crucial role in the day-to-day functioning of the College program. Whether in visiting the classes of new junior faculty or in the training of graduate students in the teaching of mathematics, Paul and Diane carefully oversee a large, high-quality operation carried on by the entire senior faculty. Thus, at Chicago the faculty is determined, in the context of a high-powered research environment, to take time and energy to fashion and nurture a first-rate college mathematics program. A consequence of these efforts is that Chicago graduates the highest percentage of mathematics majors, over 5 percent, of any highly selective U.S. university.

In addition to the high priority that the department places on excellence in undergraduate education, there is also a commitment (again rare among the highest-level research mathematics departments) to a precollegiate education. The department has been involved with precollege education for a decade and a half, starting well before such activities became politically popular and correct on the national level. This commitment is deeply engrained in, and a real point of pride for, the department as a whole. The motivation for the involvement was pure and professional. If the professional mathematicians at the top of the field do not take the initiative in improving the mathematical literacy of the high school population and the nurturing of creative minds and fertile imaginations, then who will?

Classroom at Chicago

The University of Chicago School Mathematics Project has begun to affect the shape of precollege mathematics education across the country. While Chicago School Mathematics efforts started many decades ago, the current project grew out of the work of Professors Paul Sally, Zalman Usiskin, Max Bell, and Izaak Wirszup. Beginning in 1983 with funding from the Amoco Foundation, UCSMP has developed a series of mathematics textbooks for K–12 and sponsors conferences and teacher development programs. There are now about three million students using UCSMP curricula. A central goal of the project is to "upgrade the mathematics experience of the average student." This goal is approached in a variety of ways: by examining mathematics curricula from the rest of the world, by looking closely at the mathematical skills that students actually bring to the classroom, and by removing the two-year pause that often separated sixth grade arithmetic from ninth-grade algebra.

The Young Scholars Program was begun in 1988 by Paul Sally and Diane Herrmann. It is aimed at students in the Chicago public schools, specifically, at the best one or two students in each school. A hundred students about to enter grades 7 through 12 come to the University every day for four weeks. Mornings are devoted to classes taught by mathematicians (in topics like geometry, number theory, coding theory, and computers and chaos). During the afternoons, counselors who are undergraduate students from Chicago and other universities lead small group activities, including problem-solving seminars and computer-based research. Every aspect of the program emphasizes a variety of career paths related to mathematics. There are weekly discussions with people from inside and outside the University whose work involves mathematics: astronomers, actuaries, engineers, physicists, computer scientists, and mathematics teachers, among others. For the older students, an admissions officer from Chicago makes a presentation about how to find an appropriate college or university. The program has helped to make the University of Chicago visible and accessible to many students who might otherwise never have considered it. More than ten alumni have become mathematics majors at Chicago, and eight or nine alumni enter the university each year. The program costs about $100,000 for one hundred students, all of whom commute from home. Support comes from the National Science Foundation, the Office of Gifted Programs of the Chicago Public Schools, and from the University of Chicago Mathematics Department. Paul Sally believes that a similar program could be run for as few as fifteen students. A ratio of four students to one counselor is good, but six to one is possible.

In 1991 Paul Sally organized Seminars for Elementary Specialists and Mathematics Education, or SESAME. Classes for fourth- through eighth-grade teachers are taught by faculty from the University of Chicago, Northwestern, the University of Illinois at Chicago, and other universities. The goal is "to develop a deep understanding of the conceptual foundations of mathematics, to generate activities that students can use to explore abstract ideas, and to convey a sense of mathematics that evolves from the ideas that are presented in an elementary school classroom." Classes meet for three hours on ten Wednesday afternoons from January to June and for six hours a day during a two-week summer program. Participating teachers receive academic credit toward state endorsement as mathematics specialists; this follows three years of participation, or 270 contact hours. During this time they may take eight or nine courses on topics such as "Geometry with applications to the elementary school classroom", "Probability and statistics with applications to the elementary school classroom", and so on. Lectures are extremely interactive. The Chicago program reaches more than a hundred teachers a year, at a cost of $2,000 per teacher. Sally believes similar programs ought to be widespread. A reasonable scale to begin with is ten or twelve schools and two or three teachers from each school.

Robert Fefferman has continued a program begun by Israel Herstein for Chicago high school mathematics teachers and students. There are sixty-five high schools in Chicago, of which about ten offer a calculus course. The program involves twenty high school teachers, each of whom brings a student, and they take an analysis course together. This program has received much praise and has attracted an exceptionally strong endorsement from the University president, who sees this as yet another example of the positive leadership role played by the University in the city community.

There is significant interaction among the department's outreach activities. The SESAME program is based on curricular ideas developed by UCSMP. Counselors for the Young Scholars Program are often drawn from a Summer Research Opportunities Program, which brings undergraduate mathematics majors from historically black colleges to Chicago. While many of these programs began with funding from government sources, they are also the kind of programs that should attract financial resources and partnerships from industry and the private sector.

Financial Mathematics M.Sc. Degree Program

Designed to take advantage of the increasing use of mathematics in the field of finance, this program has just been introduced by the Department of Mathematics at the University of Chicago. The goal is to produce graduates who understand the theoretical backgrounds underpinning various models used in the financial markets for pricing, hedging, assessment of risk, etc. The course consists of a combination of basic mathematics (thirty weeks with three lectures per week on numerical methods, differential equations, neural nets), probability theory (twenty weeks with three lectures per week on stochastic calculus), and economics (ten weeks with three lectures per week on the economics of uncertainty and capital and pricing), with lectures from experts in practical applications (thirty weeks with four lectures per week on simple option models, portfolio theory, fixed-income derivatives, foreign exchange, advanced option pricing, and risk management). The first three sections are taught by faculty from the mathematics, statistics, and economics departments respectively. The later sections and the applications are taught by experts drawn from a cross-section of the Chicago financial world.

The program seeks students who have a strong mathematics and/or science background who are interested in financial applications and in making a career in this area. It is a program grounded in mathematics and mathematical thinking rather than a program designed to teach a few mathematical tools to people with a background in economics and finance. It is expected that graduates will have a sufficiently strong background to adapt their models to changing market circumstances. Although it is too early to declare success, the program has started well, with 28 full-time equivalent students, of which 23 are taking the program full-time. It was begun with a loan given to the Department of Mathematics by the University. It was initiated by the department, who saw opportunities to fill a real need and to become involved in a revenue-producing operation. The department receives a certain fraction of the $27,000 tuition for each student and funds to cover the regular departmental responsibilities of the two and a half faculty members who run and teach in the program.

Although the program relies on the strong reputation in mathematics and economics enjoyed by the University of Chicago and, as currently organized, takes advantage of a local pool of talented colleagues in the financial world who can teach applications, it should serve as both a model and a stimulus for other mathematics departments to consider ways in which they may generate revenue and develop interfaces with important areas of application. It is also worth stating again that this is a program that is led by mathematicians, emphasizes mathematics and mathematical thinking, and takes advantage of the increasing reputation of mathematics in the economics and finance worlds.

It was certainly a grand and novel experience to see mathematicians leading ventures rather than simply providing peripheral support. In the past year several other universities have followed Chicago's lead into the interface of mathematics and finance.

Chapter 11
University of Arizona[1]

The Task Force team of Mort Lowengrub, Carl Cowen, John Garnett, Jim Lewis, and Raquel Storti visited the University of Arizona Mathematics Department February 27 and 28, 1997.

Arizona was selected for a site-visit for two reasons. First, the Department has a heavy service teaching responsibility. Finding out how it met that responsibility was interesting. Second, the Department has deliberately focused on certain areas of endeavor, notably mathematics education and applied mathematics. (It should be noted that the former chair of this department, Alan Newell, is a member of our Task Force.)

	1995–1997 Average /yr
Students	
Full-Time Undergraduates	21,511
Junior/Senior Majors	230
Master's Degrees Awarded	8
Ph.D. Degrees Awarded	3
Full-Time Graduate Students	58
First-Year Graduate Students	21
Fall Term Course Enrollments	
Below Calculus	2,872 (44%)
First-Year Calculus	1,975 (30%)
Other Undergraduate Courses	1,382 (21%)
All Undergraduate Courses	6,229 (95%)
All Graduate Courses	314 (5%)
Teaching Faculty	
Full-Time Tenured or T-track	59
Full-Time Non-tenure-track	32
Part-Time	3

The visitors received an enthusiastic and gracious reception from the Arizona faculty, staff, and administration. Arizona is an excellent example of a mathematics department that understands its role within its university and performs this role with distinction. University administrators laud the department's concern for students, particularly for students from other majors; its interest in teaching innovation and in quality teaching in general; and its atmosphere of cooperation between mathematicians and mathematics educators.

[1] Number of full-time undergraduates in table is taken from the National Center for Education Statistics, U.S. Department of Education, Fall Enrollment, 1996. Other data is from the AMS-IMS-MAA Annual Survey, 1995, 1996, and 1997. Course enrollment figures are from fall 1998.

Faculty from other departments praise the Arizona mathematicians for their pleasant accessibility and eager collaborations.

The department consciously tries to do some things very well and to expend minimal effort on activities it does not think it can do well. Besides mathematical education and applied mathematics, things the department does well includes innovative teaching of small classes and interaction with the Arizona high schools.

This report touches on seven aspects of the Arizona department:

Entry Level Courses

The Teaching Environment

Temporary Faculty

Mathematics Education

The Mathematics Center

The University-School Cooperative Teaching Program

Applied Mathematics

Entry-Level Courses

The Arizona department has a large service course load. In the fall semester 1998 there were over 3,100 enrollments in 93 sections of four different below calculus-level courses and finite mathematics, and nearly 2,800 enrollments in 81 sections of five different calculus courses. The large variance in the mathematical preparation of Arizona freshmen makes the department's service course job more challenging.

For many years the University of Arizona had inadequately supported precalculus teaching, until by 1984 resources had fallen to the point that College Algebra was offered in classes of over 100 students and in a self-study program of 5,000 students, while Finite Mathematics and Business Calculus were taught in classes of 300 to 600 students. Fewer than 55 percent of the students enrolled would complete these courses with a passing grade. In response to external and internal reviews, the department presented the University administration with a "Decision Package" in 1984. The package proposed a required mathematics placement test for all freshmen and class sizes of at most 35 in all courses except Business Calculus. In exchange the department promised to provide University of Arizona students "first-class mathematical instruction," to solve the problem of high attrition and failure in entry-level mathematics courses, and to upgrade the mathematics backgrounds of secondary school teachers.

The proposed package received administrative support in the concrete form of 10 new permanent faculty positions and an eventual annual budget supplement of $800,000 for visiting faculty. The plan to reduce class sizes was carried out on schedule, and by fall 1998 Arizona course listings included 50 sections of College Algebra, 12 sections of Trigonometry, 16 sections of Finite Mathematics, and 60 sections of Calculus I, II, and III. Most happily, the shift to small classes coincided with a dramatic and demonstrable improvement in student performance: between 1985 and 1990, the passing rate in undergraduate mathematics courses jumped from 55% to 77%, while enrollments in mathematics classes increased by nearly 30%.

The Teaching Environment

However, even small classes must be taught well, and at Arizona they are taught well. The department has a long-fostered environment in which teaching and research are of equal importance. A solid TA training program has been in place since 1985. The department offers regular seminars and workshops in which ladder faculty, lecturers, and teaching assistants learn innovative teaching methods. The research faculty treats the teaching faculty and the teaching assistants as equal colleagues and encourages them to develop new course materials. By departmental policy, every faculty member routinely teaches incoming freshmen. On several occasions, faculty promotions and salary increases have been justified completely on pedagogical contributions.

Temporary Faculty

Teaching 93 sections of precalculus and finite mathematics costs money, and the University of Arizona is not rich. Some of the courses are taught by regular faculty and some by teaching assistants, but most are in the hands of instructors

The Arizona Math Building

or lecturers. In 1998–99 the department employed 20 FTE's as instructors or lecturers on contracts ranging from one to three years, of whom nine were full-time lecturers on multiyear contracts. (By University policy, individuals who have a half-time appointment or higher for an academic year have the same benefits available to regular faculty, including health and retirement. Their offices are equipped with one or more computers, which are connected to the departmental network and the Internet.) The typical teaching load of a full-time instructor/lecturer is three courses per semester. The current starting salary of an instructor/lecturer is $25,000, and individuals on multiyear contracts earn between

$30,000 and $40,000. In addition to the instructors and lecturers, the department has visiting faculty from the high schools and the local community college. In 1998–99 there were four postdoctoral faculty and six visiting faculty. While it is undesirable to assign so many University courses to temporary faculty, the department has no other way to teach its beginning courses in small classes. They cannot afford the 40 new ladder faculty needed to teach 160 classes, and they cannot accommodate the 80 additional graduate students needed to cover 160 TA sections. The department recently established a three-year mathematics teaching postdoctoral position for recent Ph.D.'s in mathematics or mathematics education. The starting salary for teaching postdocs is currently $35,000 per year for a teaching load of two or three courses per semester, including upper-division courses. They are also provided with a small professional travel allowance. The department strongly encourages teaching postdocs to develop both their research program and activities in pedagogy and curriculum reform under the mentoring of a faculty member. The intention is that three years in the Arizona teaching environment will prepare them for future academic employment at four-year or master's institutions.

Mathematics Education

The department has instituted a Ph.D. program in mathematics education that requires 36 units of graduate mathematics courses, including algebra, real analysis, geometry and topology, and the same qualifying examination schedule as mathematics Ph.D. students. Students are also required to have at least two years of precollege teaching experience. Mathematics education Ph.D. theses entail research in mathematics education or the history of mathematics. Mathematics education grants account for 40% of the department's external funding. In the department there is a genuine spirit of cooperation between faculty in mathematics education and mathematics itself.

The department's three highest nonadministrative faculty salaries belong to professors in mathematics education. To encourage the equality between basic research, teaching, and educational research, the Faculty of Science (since renamed the College of Science) in 1992 established the Science Education Promotion and Tenure Committee, which provides a separate advancement track for faculty interested in precollege mathematics or science education.

The Mathematics Center

The Mathematics Center is a drop-in advising center for undergraduates considering a mathematics major. It provides students with quick answers to technical advising questions, and refers students with more academic questions to one of the faculty advisors. The Center has an undergraduate lounge and a small undergraduate library. It has instituted a series of undergraduate mathematics colloquia, it publishes a newsletter, and it sponsors career days and "math movies". The Center has instituted drop-in tutoring for upper-division mathematics courses, and it has been instrumental in the creation of the annual $1,000 Outstanding Mathematics Advisor Award.

The University-School Cooperative Teaching Program

In 1998–99 the Cooperative Teaching Program brought six teachers from local high schools or community colleges to the department, while the department sent their institutions six replacement teachers, some of whom were recent graduates in mathematics education. The University pays the replacement teachers $25,000 per year, and the school or college continues to cover its own teacher's salary and benefits. Visiting teachers teach four or five semester courses, take four advanced courses, participate in the Mathematics Instruction Colloquium and do a research project that will help the teacher's school district. The resulting communication between University of Arizona and the schools is beneficial for the University, for the schools, and for their students. Professor Elias Toubassi must be credited for this excellent ongoing program.

Classroom at Arizona

The Program in Applied Mathematics

Since its inception in 1978 the interdisciplinary program in applied mathematics has been oriented towards nonlinear analysis and computer simulations. Research topics have included shock waves, laser optics, pattern formation, turbulence, soil mechanics, kinetic theory, the earth's core, integrable systems, and population dynamics. After a fairly standard set of first-year courses (all taught by mathematics faculty), each student has an individual program that may involve problems from biology and image reconstruction to numerical PDE. One unique feature is Director Michael Tabor's Applied Mathematics Laboratory, where, in a one-year course, students make actual experimental observations in order to see firsthand how a modeling problem comes about. About 50 percent of the applied mathematics Ph.D. theses have been directed by mathematics faculty. (Note: The Arizona applied mathematics program is also discussed in the "Interdisciplinary Section" of Chapter 13.)

The following document explains the department's promotion and tenure criteria. Because it plays a large role in much of what is discussed above, we have included this document to illustrate how one department handles such matters.

An Overview of Performance Criteria for Promotion and Tenure in Mathematics

The purpose of this document is to give a larger perspective of the criteria the Promotion and Tenure Committee employs in arriving at its recommendations. These criteria are consistent with the stated guidelines of both the Department of Mathematics and the Faculty of Science, as well as the standards used by the mathematics community in general, and by some of the "top ten" departments in particular. In doing this, we wish to highlight some of the peculiarities of the mathematics community that set it apart from other academic disciplines.

In accordance with the land-grant charter of this University, we consider the contributions of each of our colleagues to the creative, the instructional, and the service missions of the University. We insist that quality be achieved in all these areas and that this work must be at a level consistent with our departmental goal of being one of the top ten departments in the country, i.e., the candidate must compare favorably with peers at that level.

A. Judging Stature and Excellence in Creative Activity

For most faculty members of the Department of Mathematics, creative activity constitutes research in mathematics. However, there is a significant fraction for whom creative activity constitutes research in mathematics education. The procedure for an individual to identify with one group or the other is clearly spelled out in the promotion and tenure guidelines of the department, and the Committee treats each case accordingly, as prescribed in the Faculty of Science guidelines. In both cases all the criteria used to judge creative activity are based on peer review, either direct or indirect, thereby indicating the regard and respect in which the candidate is held in his or her field.

It is absolutely essential that the Committee be able to ascertain that the candidate's work is of real significance, of high-quality, and sustainable. As a matter of practice, this judgement is invariably influenced by informal interaction, seminar presentations and the like, but as a matter of principle the judgement should ultimately be based on a real understanding of the candidate's work. Specifically, this understanding is built upon consideration of solicited letters from referees, the publication record, grants and awards received, as well as other indications of professional distinction.

A.1) Letters from Referees:

This is the crucial measure. The letters must indicate that the candidate's accomplishments are well known and highly regarded by the acknowledged experts in the candidate's field or fields. In both mathematics and mathematics education, most research is conducted individually or in small groups (as opposed to teams). One consequence of our standard, and of the individual nature of this research, is that the candidate must be consistently judged by the referees to have done significant independent and original work, demonstrating an ability to pick and solve problems of interest. It is not sufficient to have shown great zeal at extending the ideas of others. The referees often give insights into the publication record of the candidate, including his or her relative contribution to collaborative work.

A.2) The Publication Record:

It is impossible to describe in a uniform way how we actually identify work of high-quality. However, it is easy to identify an important caveat, for in mathematics quality is not always correlated with quantity. For instance, two of the premiere number theorists of this century, Artin and Hecke, each published fewer than fifty papers throughout their long careers. This value of quality is reflected in the fact that the Mathematics Division of the NSF limits publication lists to ten and that many mathematics departments, such as Harvard's, base tenure decisions on a candidate's five best papers. It is also important to emphasize that standards within mathematics are neither homogeneous nor static. Applied mathematics is closely akin to scientific disciplines, often even having an experimental component. As in most sciences, the formulation of a problem in applied mathematics can sometimes undergo a long and tortured evolution, involving many false or incomplete steps, and it is important for a researcher to leave his or her imprint along the way (pointing in the right direction, of course). In contrast, pure mathematics is not a scientific discipline at all. Results in pure mathematics consist of mathematically rigorous solutions of precisely stated problems. Such results are stated as "theorems", the demonstrations (proofs) of which are either correct or not and, once established, are not subject to change upon reexamination (although it is highly regarded to discover a major simplification in a long proof).

In many areas of both pure and applied mathematics, there is a strong consensus that emphasis should be placed on publishing rigorous and complete papers. A short "four-page" announcement of results with a loose outline of the arguments is rarely considered as a significant work, even if it appears in a refereed journal. This is because there is an enormous difference between seeing a reasonable strategy for a proof and actually carrying out a proof. The resulting emphasis on completeness sometimes dramatically slows the publication process and is reflected by two phenomena: (1) mathematicians often circulate their papers as preprints for extended periods prior to submission for publication, (2) the delay in publication for prestigious mathematics journals is frequently at least two years. For these reasons referees will often be familiar with and comment on works that have not yet appeared in journals (although these works have usually been submitted). Mathematics is slow and difficult, and there are sound reasons for emphasizing reliability over quantity. We should also note that in mathematics the order of authors' names on an article is usually alphabetical, with no regard to seniority or percentage of contribution. One seldom sees more than three authors on a mathematical paper, especially in pure mathematics, and usually all authors make essential contributions to the work. As a result, if asked to do so, candidates will often simply divide the percentage of contribution evenly among the authors. These circumstances sometimes muddle the issue of which author or authors are responsible for the key ideas; however, this matter is occasionally addressed by the referees, and we often have some knowledge of our own. We try to make this explicit in the individual reports.

A.3) Grants and Awards:

Opportunities for funding within mathematics vary greatly from field to field. Government agencies such as the Air Force, Army, Navy, Department of Energy, and National Institutes of Health usually make a small part of their overall research budgets available to applied mathematicians. However, pure mathematicians have far fewer sources of funds from which to draw. Even when funding is available in mathematics, the award sizes are much smaller than those in other scientific fields. For instance, the median annual award sizes for NSF grants from the Mathematics and Physical Sciences directorate (MPS) in disciplines other than math (astronomy, chemistry, materials research, and physics) are two and a half to almost four times as large as those in mathematics. The median annual size of NSF grants in mathematics was $22,862 in 1996 and $28,000 in 1997. In light of this situation, it is to be expected that candidates in mathematics will have funding levels substantially below candidates from other disciplines. However, when compared to other mathematics departments, the mathematics department at the University of Arizona does well. In 1997 it ranked twentieth among universities in terms of funding received from the NSF in mathematics.

Members of the Committee regard awards, such as Sloan Fellowships, as absolutely reliable indicators of the quality of the candidate's work, and outside support as a valuable indicator of its impact and potential. However, in mathematics neither is regarded as essential. There are three reasons for this. First, the great bulk of outside support comes from a single source, the NSF. Second, NSF funding for mathematical research has been and continues to be very tight. Third, in most areas of mathematics, a lack of outside support is not a real hindrance to continued productivity in research. The situation in mathematics is not comparable to that in other sciences where support for a research lab and assistants is essential. We are not aware of any mathematics department in the country that insists on outside support.

A.4) Postdoctoral Positions:

At variance with most scientific disciplines, in mathematics a postdoctoral position is often considered to be a prestigious award for a pretenured faculty member. This is the case for positions that allow researchers to conduct their own research, often with a reduced teaching load. These may come under the guise of "named instructorships", such as the Moore Instructorships at MIT, the Miller Fellowships at Berkeley, or our own Pierce and Rund Instructorships. They may also be postdoctoral fellowships awarded by a major research institute like the Mathematical Sciences Research Institute (MSRI) in Berkeley, the Institute for Mathematics and its Applications (IMA) in Minneapolis, or the Institute for Advanced Study (IAS) in Princeton. They may also be postdoctoral fellowships awarded by a government or private funding agency. In any guise, there are very few of these positions.

The Mathematical Sciences Postdoctoral Research Fellowships awarded by the NSF each year provide a notable case in point. Despite the name, these fellowships are quite different from traditional postdoctoral positions in the sciences. First, the research plan for each fellowship is prepared by the applicant, not by a mathematician at the host institution. Second, each of these fellowships is awarded directly to the applicant by an NSF panel of mathematical scientists, not by a senior researcher at the host institution. Third, only 25–30 of these fellowships are awarded each year. Recipients of these fellowships often have concurrent tenure-track appointments. They also typically teach. For these reasons, the Committee members regard these awards as quite prestigious and regard the years of fellowship as time spent at the assistant professor level.

A.5) Other Measures:

There are other measures used by the Committee, such as invitations to speak at conferences, contributions to conference proceedings, and seminar participation. While we consider such indicators positively, less weight is attached to a lack in the first two measures than in other disciplines because in many areas of mathematics there are not as many conferences as in, for example, areas of Physics. This reflects the general funding situation of mathematics. For similar reasons, memberships on professional committees, while a solid indicator that the candidate is regarded highly within the mathematics community, are not considered to be essential (being fewer per capita than in other disciplines).

B. Judging Excellence in Teaching

The criteria used by the Committee are essentially those spelled out in Section II of the Faculty of Science Statement on Guidelines.

In evaluating teaching we try to take a balanced approach in weighing student evaluation forms, student comments, peer review, and follow-up interviews with students. The Committee expects faculty to set high standards in all courses, but pays particular attention to teaching performance in lower-division undergraduate courses. We also value contributions to all instructional programs through resource development (like new courses, textbooks, and software), especially if it is nationally recognized through grants or awards. Involvement in undergraduate advising, the Honors Program, minority mentoring, graduate admissions and advising, the preparation and grading of qualifying and preliminary written exams, and oral exam committees are also contributions that weigh in favor of a candidate. In addition, the Committee fully recognizes similar contributions by faculty to interdisciplinary programs.

It is expected that faculty attaining the rank of full professor will have played a significant role in developing the quality of our graduate program. Perhaps the most direct way of contributing to this is to guide the research of a student in the Ph.D. or master's programs. However, this is not the only way. Organizing and participating in working groups or seminars which introduce students to current trends or research problems is a valued activity in a department as diverse and interactive as ours. Of equal importance is participation in the design and teaching of the core graduate curriculum as well as the more specialized year courses designed to lead students into active research.

In mathematics it is rare that a pretenured faculty member takes on a Ph.D. student. What we look for in such faculty members are the qualities that make for an effective adviser, such as the ability to interact one-on-one with students. The input on such judgements is generated by comments and letters from graduate students and our own observations.

C. Judging Excellence in Academic and Cultural Service

There is an important point to be made here. The Mathematics Department does a great deal more teaching than the other departments in the Faculty of Science. For this reason it is quite natural that our academic service contributions tend to be internal to the department and University and are often teaching oriented. Academic service external to the University includes: refereeing for, sitting on editorial boards of, and serving as editor of professional journals; serving on national organizing or governing committees of, or holding office in professional societies; serving on peer review panels; organizing programs at or serving on the governing boards of research institutes.

Because of the centrality of mathematics to primary and secondary education, a large component of our cultural service is comprised of programs with local school systems, designed to enrich their students and faculty. Several such programs specifically target local minority populations. The quality of such programs is often reflected by their recognition through grants and awards.

In mathematics we do not expect extensive service contributions from pretenured faculty, although we do look for journal refereeing activity. We expect pretenured faculty to concentrate principally on teaching and research. However, for promotion to full professor we expect an extensive and productive service record, both external and internal to the University.

Chapter 12
University of Texas at Austin[1]

The Task Force visited the University of Texas at Austin on December 4–5, 1996. Representing the Task Force were Carl Cowen, Douglas Lind, Mort Lowengrub, Don McClure, and Raquel Storti. This report focuses on the Emerging Scholars Program at the UT Mathematics Department, which has been successfully copied at scores of institutions. The report also discusses the department's Actuarial Studies Program, Introduction to Research Lectures for graduate students, and Saturday Morning Math Group for local high school students.

	1995–1997 Average /yr
Students	
Full-Time Undergraduates	31,121
Junior/Senior Majors	335
Master's Degrees Awarded	26
Ph.D. Degrees Awarded	12
Full-Time Graduate Students	104
First-Year Graduate Students	19
Fall Term Course Enrollments	
Below Calculus	2,397 (23%)
First-Year Calculus	4,835 (46%)
Other Undergraduate Courses	3,082 (29%)
All Undergraduate Courses	10,314 (98%)
All Graduate Courses	263 (2%)
Teaching Faculty	
Full-Time Tenured or T-track	65
Full-Time Non-tenure-track	13
Part-Time	10

Emerging Scholars Program

The Emerging Scholars Program (ESP) is a joint project of the Department of Mathematics, the Charles A. Dana Center, and the College of Natural Sciences. It aims to stimulate and assure the success of highly qualified but "at risk" students in freshman calculus. The program's immediate goal is to increase the numbers of women, underrepresented minorities, and rural white males who

[1] Number of full-time undergraduates in table is taken from the National Center for Education Statistics, U.S. Department of Education, Fall Enrollment, 1996. The remaining data is from the AMS-IMS-MAA Annual Survey, 1995, 1996, and 1997. The table reports the average of all available data provided by the department during the three-year period.

excel in calculus. A longer-term goal is to develop minority and women mathematicians and scientists. A distinguishing feature is providing ESP students with a mathematically rich and challenging environment combined with a community life focused on shared intellectual interests and professional goals.

The heart of the ESP lies in the intensive discussion sections attached to a standard calculus course. These sections, limited to 24 students each, meet three times a week for two hours (while students in standard discussion sections of 40 students meet twice a week for one hour). For this extra work an ESP student earns two pass/fail credit hours in addition to the four credit hours all regular calculus students receive. Each ESP section is led by an advanced graduate student together with one or two undergraduates who are ESP alumni. Students work individually or in groups on carefully crafted problems ranging from average difficulty to those designed to stimulate independent thought and group discussion. The Task Force representatives observed one ESP section and participated in the discussions with students. The environment was very lively, and the group interactions were highly effective. For one Task Force member this experience was the most striking and memorable of the entire work of the Task Force.

Students must meet academic standards to be eligible for the ESP. The standards include aptitude gauged by SAT scores and achievement gauged by class rank in high school and performance in previous math courses. Students understand that they will be asked to work more and that they will develop ability to work independently on problems that go beyond routine ones. Students become a part of a peer group of highly motivated and equally capable students. In the friendly, supportive environment of the ESP section, they work together to share insights, learn from each other, and experience using the course material for meaningful problems.

While a small segment of the discussion section may involve presentation of material or concepts related to the regular calculus lectures, the focus of the sections is on group problem solving. Students work in small groups of 4 to 6 solving problems posed by the advanced graduate student associate instructor (AI) on a section "worksheet". The section staff are available for support, and they help guide group discussions, but the students learn to be self-reliant in solving the problems. One of the principal creative tasks of the AI is the careful design of the worksheet to include suitably difficult and interesting problems. The prior training of the AI stresses the importance of this role.

The ESP program began in the fall of 1988, under the leadership of Efraim Armendariz, with one section of 21 students. By the time of the site-visit there were seven sections with a total of about 120 students. The ESP was adapted from the Professional Development Program Mathematics Workshop developed by Uri Treisman at UC Berkeley. In 1991 Treisman was hired by UT Austin, and he brought with him the Charles A. Dana Center. Treisman was named a MacArthur Fellow in 1992 in recognition of his creativity in education.

The ESP has a marginal cost per year of about $120,000 or marginal cost per student of about $1,000. This includes stipends of about $12,000 each for six graduate student teaching assistants, program "overhead" of $30,000 for a coordinator responsible for administration and recruitment, and $20,000 for other costs such as staff training.

The program has compiled detailed and compelling data about its successes in better performance and higher retention rates. For example, since the start of the program ESP students have typically earned grades one-half to one full grade point higher than the class average. These data have convinced the UT admini-

The Texas Math Building

stration that the program works and is well worth the costs. We found uniform and enthusiastic support for ESP from the administrators with whom we met.

One of Triesman's key observations was that a primary cause of students' poor performance was academic isolation. The group work in ESP discussion sections counteracts this isolation and helps form social bonds that typically last far beyond the end of the course. Our observation of an ESP section showed that group conversation mixed serious work on the problems at hand with what's happening next weekend, with comparing notes about a physics course, and so on—weaving academic accomplishment into each student's life.

Although ESP students work longer and harder than other calculus students, they respond very well because they feel part of a cooperative effort. Here are some sample comments from students:

"By brainstorming and working together, we figure out what's going on and are able to handle problems from the homework on our own."

"From this program some of us have realized our potential and know that as a whole, we can boost each other's potential."

"I was in Emerging Scholars the first semester, but I decided to leave the second semester because my course was too heavy. After two weeks in the second semester of calculus, I decided to return to the program. I just did not realize what an asset the program was. The extra hours of work allow us to better comprehend the topics introduced in class."

It should be noted that the ESP program's creation of small learning communities helps significantly with problems of transition when moving from high school to a large state university.

A vigorous recruitment effort is run by the ESP coordinator. In the spring prior to matriculation, newly admitted students are screened on the basis of SAT scores, class rank, and previous math achievement. The eligible students are "nominated" for the program and offered an opportunity to participate. The invitation letter stresses the "honors" aspect, the extra work involved, and the higher expectations for level of attainment in the calculus course. The screening effort especially seeks minority students and students from small towns; it also tries to attain some gender balance. Singling out students with letters of invitation works very well: students say they enroll in ESP "because they were asked."

During freshman orientation, all students intending to take calculus are informed about the ESP and the entrance requirements. At this point every eligible student has an opportunity to express his or her interest.

The ESP is selective but not exclusionary. While the program strongly encourages participation of women and underrepresented minority groups, it is open to all students who meet the academic selection criteria. All students who meet the selection criteria and choose to participate are accommodated. At the time of our visit to UT, the program was serving about 120 calculus students out of an estimated 1,500 students in the regular calculus sequence. About two-thirds of the students are African American, Hispanic, or Native American. The rest are white non-Hispanic or of Asian ancestry. About half of the participants are women.

The ESP model is highly adaptable and has been successfully used around the country. See Chapter 13 for a description of full and partial adaptations of ESP at other doctoral institutions. Texas runs training sessions for implementers elsewhere.

ESP is designed not to intrude on the normal calculus instruction, but to be an added enrichment program, so disruption to a department is minimal. However, in the UT mathematics department ESP ideas have spread to many faculty's approach to teaching. In all likelihood, the combined efforts of ESP and the Actuarial Studies Program have led to a large increase in the number of mathematics majors from underrepresented groups. In 1997, for example, the department had 404 majors, of whom 185 (45.7%) were women and 103 (25.4%) were African American or Hispanic (46 in the intersection). Within the University as a whole in 1997, the undergraduate enrollment was 33,800, of whom 16,805 (49.7%) were women and 5,316 (15.7%) were African American or Hispanic.

Actuarial Studies Program

A very small actuarial program was started at the University of Texas in 1913 in the Mathematics Department and moved in 1958 to the Finance Department in the Business School. In 1988 when the Finance Department canceled the actuarial program to focus its resources on mainstream finance, the Texas actuarial community approached the Mathematics Department to take back the program. The department agreed in principle to take on the program, providing one faculty member and one TA to teach full time in the program, while the actuarial community agreed to raise several thousand dollars a year for operating expenses. The first challenge was finding a faculty member to teach actuarial science and run the program.

James Daniel, the chair of the Undergraduate Curriculum Committee at the time and former department chair, was ready for a new challenge and eagerly accepted the responsibility to run the program.

Daniel teaches five or six actuarial science courses, four each year. The rest of the actuarial studies program consists of existing mathematics courses in calculus, probability and statistics, etc., plus selected courses in business, accounting, and finance. Instead of learning just the material needed to teach core courses, Daniel wanted to become deeply knowledgeable about actuarial science. He spent one year of intense study and managed to pass all the actuarial exams required for associate membership in the Society of Actuaries.

The Texas Math Computer Lab

Under his guidance the Texas actuarial program has flourished. It graduates close to 20 students a year, and most quickly have multiple job offers. Daniel is in his office ten hours a day recruiting students, advising them and later helping place them, championing his program among the Texas actuarial community (for good will, internships, and fund raising), working with alums, and more. The Texas actuarial community now contributes over $40,000 in expendable annual gift funds for a mixture of scholarships and operating expenses (e.g., printing and mailing résumé booklets and an alumni newsletter). During 1994–96 the insurance industry raised $300,000 to triple a modest endowment of an actuarial science professorship that Daniel holds; and the university added more than $150,000 in matching funds. Daniel has organized a national caucus of mathematics faculty interested in actuarial science who meet regularly at annual AMS/MAA national meetings.

Introduction to Research

The department sponsors a special series of outside speakers each year, primarily intended to introduce graduate students in their first two years to different areas of research. At the time of the site-visit, this program, called "Introduction to Research". had been running for three years. The focused speaker program was started by Dan Freed and John Luecke and was funded initially through matching funds associated with NSF PYI awards.

Approximately six scientists participate each year. A committee of postdoctoral fellows, in consultation with graduate students, carefully picks the speakers based on their ability to inspire and motivate. The introductory intention of this program is made clear to invitees, as well as the expectation that they interact in a number of informal ways with graduate students during their stay. This organizational activity is regarded as part of the professional training of the postdocs. The postdocs we spoke with had a valuable sense of ownership of this program and clearly worked hard to make it a success.

Speakers have responded to the goal of the special speaker program, and attendance at the talks is high, usually more than a hundred people. The department has made a six-year commitment to support this program, at a level of about $6,000 per year from departmental discretionary funds.

Saturday Morning Math Group

At least once a month, Austin-area high school students are invited to the UT campus for a half-day program of talks and math activities guided by faculty and graduate students. The program is an outgrowth of an NSF-funded Regional Geometry Institute. Initially organized and funded by Dan Freed and Karen Uhlenbeck, the program is now supported by the Department of Mathematics. The graduate-student coordinator is responsible for identifying a topic and principal speaker, usually a local faculty member, and for developing activities and materials which engage the high school students in exploration and discovery. Attendance at sessions has grown steadily and now averages well above 100 students per session. In addition to providing inspiration and support to some of the brightest high school students in the area, the Saturday morning programs also serve to promote contacts with local high school teachers, even to the point of pulling them in to working in the teacher preparation summer programs.

Chapter 13
Examples of Successful Practices

Service Courses in the First Two Years

These courses, among which we include calculus, generate over 80 percent of the enrollments at most research mathematics departments. As discussed extensively in Part I of this book, high-quality, innovative instruction in these courses is a critical part to any plan to seek increased resources for mathematics.

The Task Force has learned about a variety of new approaches to teaching calculus. The calculus reform movement has, in combination with other factors such as cheaper, more powerful technology, had a major influence on virtually all calculus texts. A 1995 NSF-sponsored survey found that three-quarters of all doctoral mathematics departments were engaged in modest or major calculus reform. The survey documented that most "reform" calculus courses at universities were using traditional texts supplemented in recitations and labs by reform activities, such as cooperative learning and computer-based experiments. There are a dozen volumes in the MAA Notes series devoted to new technology, pedagogy, and instruction associated with calculus, e.g., MAA Notes #30, *Problems for Student Investigation.* Dozens more are available from commercial publishers. The use of technology, especially graphing calculators, in calculus instruction is now widespread; graphing calculators have been required for several years in one part of the AP Calculus exam (another part of the exam forbids them). There is now a fairly continuous range (not partially ordered) of approaches to teaching calculus.

The new approaches generally require more faculty time, but as Tennessee mathematics chair John Conway put it: Not to pay enough attention to the way calculus is taught is probably asking for trouble. The MAA Notes #39, *Calculus: The Dynamics of Change*, has a careful analysis of the additional faculty effort, resources, and other costs associated with new approaches to calculus instruction.

Nearly everyone believes that small classes are better suited than large lectures to such active learning, but to get the resources for small classes requires creative approaches by mathematics departments. The site-visit report on Michigan describes how that department turned an innovative, small-class approach to calculus instruction, which Michigan called New Wave calculus, into a vehicle for obtaining a large number of new junior faculty positions. The site-visit to Arizona documented a very different approach to keeping the size of introductory mathematics courses manageable—the use of large numbers of highly effective (non-Ph.D.) adjunct faculty. Coupled with leadership in calculus reform, the smaller class sizes have earned a very good reputation for mathematics instruction and for the mathematics department at the University of Arizona. Another good example is found at nearby Arizona State University.

Arizona State University

In the early 1990s the ASU administration identified college algebra, along with a few other courses, as a major source of discontent for students enrolled in freshman courses. After constructive discussions involving virtually all interested parties, the University and the department decided to commit substantial new resources for the establishment of a "First Year Mathematics" program within the Department of Mathematics. All courses within this program are taught in small sections using a very interactive format. Curricula were revised to reflect modern realities, and technology was integrated into almost all courses. Marilyn Carlson, a mathematics education Ph.D. from Kansas, was hired as the director of this program, and nearly thirty new lecturer positions were created. These faculty members, who typically hold three-year renewable contracts, "run" the entire FYM program in a closely coordinated manner despite huge numbers of sections in some courses—in some over 60 sections. The changes have led to dramatic improvements in grades and passing rates, including those in subsequent courses, as well as improved freshmen retention and improved general satisfaction with ASU's instructional programs in the entire community. In 1998 the FYM faculty won the coveted Governor's Award for Excellence.

It is important to note that at the same time the FYM program was being developed, the ASU mathematics department was also getting new tenure-track faculty to help advance the University's research agenda of becoming a Carnegie Research I institution, a goal achieved in 1994. The department and university leadership recognized the need to balance scholarship and instruction, and they allocated adequate resources for both.

The success of these innovations relied on proper recognition in terms of tenure and promotion, and also in salary increases. According to the department chair Rosemary Renaut, it was absolutely critical that faculty involved in instructional innovations knew that their involvement would be recognized and rewarded. The benefit to the department of its attention to its service role has been the creation of a significantly improved climate in which to request and receive new resources, whether in the form of new faculty positions or equipment for the research programs.

One method to enrich calculus instruction is with technology. Many mathematics departments have experimented with Maple or Mathematica-based calculus reform courses as an alternative to the mainstream calculus courses. The Mathematical Sciences Department at RPI (Rensselaer Polytechnic Institute) obtained a very large institutional investment in technology for its calculus course. However, there were questions about how to use this technology effectively. The concept of a studio classroom brought pedagogical changes to match the new technology in calculus. Subsequent to mathematics' use of this approach, the studio classroom became a rallying point for educational innovation across the RPI campus. Lecture, recitation, and laboratory are integrated into a single-classroom teaching style. The course TA is in the class helping the professor work with groups of students. The laboratory component involves sophisticated interactive, multimedia simulations of various physical phenomena. Students have their own laptops to enable them to take the simulations and computer algebra systems back to the dorms with them.

Precalculus reform efforts are starting to yield texts that face the difficult task of being interesting while developing skills in algebra, trigonometry, and analytic geometry needed in calculus. The title "precalculus" is a bit of a misnomer, for studies show that about one student in seven who enrolls in precalculus will successfully complete a semester of calculus. Almost twenty years ago Dartmouth started a widely copied trend of incorporating precalculus material into the first semester of calculus. The resulting precalculus/calculus course is typically two semesters long. With publishers now able to produce customized texts with chapters from precalculus and calculus texts, such combined courses are becoming easier to design.

Several innovative textbooks developed over the past dozen years have given new life to general education courses in mathematics. A number of mathematics departments have successful general education courses, in the eyes of faculty and students, using the CoMAP text *For All Practical Purposes* (also a PBS telecourse).

One of the frequent complaints by faculty is that students are inadequately prepared for freshman-level mathematics. The University of Arizona uses a novel approach to supplement placement exams. Fifteen percent of the grade in some UA freshman mathematics courses is based on a test at the end of the first week of the course that covers the prerequisites for the course. Students who do poorly have time to transfer to a lower-level mathematics course. To prepare students for these tests, admitted students are sent computer disks over the summer with samples of the placement and first-week tests along with a tutorial for learning this material.

Many mathematics departments have developed highly effective resource or learning centers to help students get more personal assistance to complement large lectures and formal problem-solving recitations. The Oklahoma State site-visit discusses their very successful resource center, which is supported by a fee paid by all students in lower-division courses. The Michigan site-visit discusses the critical role of the Michigan Math Laboratory for mathematics majors as well as students in lower-division mathematics classes. Following is a description of the mathematics learning center at Virginia Tech, which has substantially extended the services normally associated with such centers.

Virginia Tech

Recently, the provost at Virginia Tech has been touting to other provosts the exciting Mathematics Emporium that was created at his institution. That sort of publicity is rare for mathematics instruction. The Emporium, opened in fall 1997, is located in a formerly empty building near the Virginia Tech campus. It is open 24 hours a day, seven days a week. Faculty, assisted by a large number of graduate students and undergraduates, are present 14 hours a day for help. There are 500 computers grouped attractively in study pods of six. There is 24-hour technical help with the computers. At the pods and elsewhere in the Emporium much of the space is laid out to facilitate students working in groups. Other areas are designed for individual study.

The idea of teaching math in a computer-enhanced environment got its start in the spring of 1993 when the mathematics department began using Mathematica in two of its first-year calculus courses. With nearly 2,000 students taking the new classes each semester, convincing assessments were possible that showed that students in the new Information Technology (IT) initiative had final grades that were half a grade higher than those of students in traditional courses (both groups of students had a common final exam). Later assessments showed that IT students taking other mathematics or engineering courses were doing better than students who had been in traditional math classes.

"By the spring of 1995, other colleges were encouraging the department to bring all its lower-level courses into the stream of technological change," mathematics chair Robert Olin recalled. The faculty changed the precalculus course, with similar positive results. As the department continued to reform its lower-level courses technologically, the concept of the Mathematics Emporium was born.

The computers give students extensive diagnostic quizzes, electronic hyperlinked textbooks, and interactive, self-paced tutorials. All the software at the Mathematics Emporium can also be accessed over the Internet. The risk for alienation of students through computer-driven instruction is turned instead into an opportunity to spot and correct problems before they become critical. The instant feedback that computers can give maintains students' attention much better than do homework assignments that are returned days after they were done.

The Emporium environment for learning basic course skills enabled the department to redesign the traditional classroom component of some courses. In one mathematics course, classes are now grouped by major field, and more time is dedicated to showing examples connected to individual interests. Many introductory courses have course testing done in the Emporium, freeing up valuable class time. Moreover, the Emporium setup allows professors to give tests in multiple versions that students can take when they are ready. In some cases, tests can be taken more than once, in the spirit of the Keller plan (popular in the 1970s).

Interested readers can consult the Virginia Tech Mathematics Emporium's home page at http://www.emporium.vt.edu.

Successful Undergraduate Mathematics Majors

The major challenge for most mathematics departments is attracting significant numbers of students to major in mathematics. Unfortunately, the broad appeal of mathematics of the 1960s is gone, when 5 percent of entering freshmen wanted to be mathematics majors and the mathematics department could set high standards to wean the percentage down to about 2 percent. For almost three decades, less than 1 percent of entering freshmen have expressed an interest in majoring in mathematics. Several mathematics departments have successfully countered this trend by broadening the constituency for the mathematics major with alternatives to the standard mathematics major. However, a few schools, most notably the University of Chicago, have continued to be very successful with a major program geared towards preparing students for graduate study in mathematics. Among Group I mathematics departments, Chicago has the highest percentage of graduates majoring in mathematics. In addition, Chicago has the most students going on to doctoral study in mathematics—about 50 percent of their math majors. See Chapter 10 for further information.

Inclusive mathematics major programs tend to be among the most successful in getting more majors to go to graduate school in pure mathematics than programs focused on that outcome. The inclusive programs draw in more students, many of whom are not initially thinking about graduate study. This increased number of students allows for more elective courses and generates a critical mass of enthusiastic mathematics majors on campus who make math an "in" major.

Here are profiles of three successful mathematics majors that are highly inclusive.

UCLA

For many years, the UCLA mathematics department has graduated the most mathematics majors of any U.S. university. About 175 of UCLA's 5,000 undergraduate degrees are awarded in mathematics. The department has an inclusive research tradition with very strong applied mathematics. The undergraduate program reflects that inclusive tradition with a variety of options. The graduate-study-oriented "pure mathematics" track attracts 10 percent of the majors and is like such mathematics majors at other institutions. About 25 percent of the majors are in tracks in mathematics of computation and in physical science-oriented applied mathematics, both of which are quite rigorous. The largest track, with 50 percent of the majors, is the applied science track, which has five options. It is geared to students seeking employment in business or industry after completing their undergraduate studies, although a number get an advanced degree sooner or later. Most students in the mathematics/applied science track select a decision science option: management, operations research, or actuarial science. Other tracks are preservice school mathematics teaching preparation and a joint mathematics/economics major.

The inclusiveness of the department is reflected with impressive demographics: 25 percent of the UCLA's mathematics majors are non-Asian minority students.

The department has a tradition of being a friendly place where faculty and students socialize together. To make the department a friendly place for undergraduates, it employs two enthusiastic former UCLA mathematics majors whose job is to advise undergraduates (one spends half-time managing the UCLA Mathematics and Science Scholars Program). Faculty feel these advisors are critical to the department's positive reputation among undergraduates and its large number of mathematics majors.

There are three active support groups for mathematics majors. The most important is the Undergraduate Mathematics Student Organization, which promotes:

- Academic awareness of the mathematics major
- Better student-faculty relations
- Information on career opportunities in mathematics
- A peer network for mathematics majors

Each year the UMSO runs résumé workshops, interviewing skills workshops, several career workshops and panel discussions, as well as a T-shirt contest and faculty dinners.

Bruins for Mathematics is a departmental alumni organization that provides alumni support and professional contacts for students. The UCLA Actuarial Club is for students who have an interest in the actuarial profession, and sponsors informational talks by local actuaries. The department provides review sessions to prepare undergraduate students for the first two actuarial examinations.

Vanderbilt University

The Vanderbilt mathematics department has a major that attracts a broad clientele, including many preprofessional students (premed, prelaw, and pre-MBA). A sizable number of math majors pursue a second major as well; mathematics and economics is the most popular combination. After calculus, linear algebra, and differential equations, the major may be completed with five additional courses, giving the student a minimum total of 32 hours. No additional restrictions are put on the choice of courses. Students planning advanced study in mathematics typically take much more than 32 credits and usually take several graduate courses.

Mathematics is very popular as a second major among engineering students, who typically need just four additional courses beyond those required for their B.E. degree. Applied math and statistics are the subject areas that engineers usually pursue. Including double majors, Vanderbilt has the highest percentage of its bachelor's degrees awarded to mathematics majors of any U.S. doctoral university.

The Vanderbilt program attracts substantial numbers of students, engineers and nonengineers alike to take more mathematics than they normally would. One pedagogical consequence is that the teaching opportunities for the faculty are diversified, and there is not the heavy concentration of calculus and precalculus service teaching found in most mathematics programs. Students have obviously benefited by having a stronger background and better credentials for doing graduate work or attracting more lucrative job offers.

In short, this is a program very much in tune with the liberal arts philosophy espoused by many leading private universities. Its flexibility allows and encourages students to pursue mathematics at a level and depth consistent with their career objectives.

SUNY-Stony Brook

The Stony Brook applied mathematics department offers a popular major that has much in common with both the UCLA and Vanderbilt models. Its B.S. major requires 42 credits of mathematics (and 17 credits in related departments). The major is oriented towards the "decision science" side of applied mathematics and has little emphasis on proofs. Its electives are almost all in probability/statistics or operations research.

The special strength of the Stony Brook applied mathematics program is attracting students who were good in mathematics in high school but were advised by guidance counselors to major in engineering or computer science for career purposes. Many of these students grow disenchanted with these majors and eventually turn to applied mathematics or develop a double major with applied mathematics. While only a handful of entering students express an interest in applied mathematics, about 100 of Stony Brook's 2,200 bachelor's graduates are applied mathematics majors; about half are double majors. There are also about 40 Stony Brook (pure) mathematics majors graduated each year.

There are three components of Stony Brook's success that may be applicable at other institutions. The first is curricular. The focus of the major is decision sciences oriented mathematics. Recall that the most popular option in the UCLA mathematics major is the decision science track. Many mathematics departments equate applied mathematics with topics like differential equations. Businesses today are making extensive use of statistics, operations research, and game theory to solve their problems, and generally they value people with training in these areas for positions in finance and management. Scores of Stony Brook economics majors add applied mathematics as a second major because of the strong reputation of this double major for getting into good MBA programs.

The second component of Stony Brook's success is that the applied mathematics department has very good relations with other departments through joint research and educational collaborations. There are a number of cross-listed courses and faculty with adjunct appointments in applied mathematics. This has resulted in higher-than-average mathematics requirements for other majors. For example, a computer science major who wants to switch in the junior year to applied mathematics or add it as a second major needs only five additional courses, because the CS major already requires five mathematics courses and two mathematically oriented CS courses can also be counted towards the applied mathematics major.

Finally, the department identified its two beginning junior-level courses in discrete methods and probability/statistics as key courses. Majors in computer science and engineering are required or encouraged to take these courses, which each enroll over 300 students a year. The department staffs the two courses with its best teachers and shapes their syllabi to make further study in applied mathematics look as appealing as possible.

Many successful undergraduate programs at research universities have a key person or small group who has devoted a huge amount of effort for many years to make the program succeed. At Chicago, Paul Sally and Diane Hermann have played this role for the undergraduate program. For a special component of the undergraduate program, strong leadership is absolutely essential. The University of Texas's actuarial program discussed in Chapter 12 provides a good example.

It should be noted that the attractive professional careers available to actuaries can provide an excellent basis for recruiting students into mathematics. The mathematics department at small Lebanon Valley College in eastern Pennsylvania regularly gets 10 percent of the freshman class planning on a major in actuarial science or mathematics (the national average is below 1 percent) by focusing on actuarial careers in school visits and publicity materials. While most students concerned about careers enter with actuarial plans, the majority later switch to a mathematics major.

Some of the common features of highly effective undergraduate mathematics programs that were studied in the MAA study "Models That Work: Case Studies in Effective Undergraduate Mathematics Programs" are:

- Faculty take a very personal approach to their classes, even in multisection courses
- Faculty set high expectations for students and then help them meet these expectations
- The faculty are not satisfied with the current program, no matter how successful it may be
- Placement exams are very important

St. Olaf College has an interesting way to promote the inclusive nature of its major. It has a "contract major", in which the requirements of a student's major program in mathematics are carefully thought out by a student with a faculty advisor. The negotiation of the "contract" might involve, for example, a student wanting to take an applied curriculum with virtually no proof-oriented courses beyond linear algebra, and the professor arguing for adding, say, analysis and abstract algebra; sometimes the roles are reversed, with the professor pressing for some breadth in the mathematical sciences. While individually negotiated, in reality almost all the contracts tend to follow one of three general curricula, emphasizing pure mathematics, computer science, or applied mathematics.

The approach Mt. Holyoke College has taken to inclusiveness is also worth noting. It has extended the Michigan strategy of freshman alternative entries to the standard calculus sequence. Students can start the mathematics major with one of two general education seminars—one in geometry and one in number theory (taking calculus in the junior year)—as well as a traditional or very reform version of calculus. Then all students are funneled into a sophomore Laboratory in Mathematics Experimentation, in which students explore six pure and applied open-ended problems and write 10-page papers about them. After the laboratory course, students branch out. There are some traditional courses in analysis, algebra, and geometry. Along with some traditional courses, the department also has developed topics courses with minimal prerequisites, for example, a course in knot theory and a course on Lie groups (the latter has Calculus I and linear algebra as its prerequisites).

Programs for Underrepresented Groups in Mathematics

There are two groups that historically have been underrepresented in undergraduate mathematics classes and in careers that require strong mathematical skills—women and non-Asian minorities. A number of years ago the phrase "math anxiety" was coined to refer to the learning problems, largely nonacademic, that women bring with them into mathematics classes. For a long time the problems that non-Asian minorities had with mathematics were thought to be mostly academic, that is, poor K–12 preparation to do college-level mathematics, and were addressed through special remedial programs. Then the thinking about minorities' problems started to consider other factors. This rethinking was accelerated by Uri Treisman's successful Professional Development Program (PDP) for minorities at UC-Berkeley, which achieved remarkable successes by setting high standards and emphasizing collaborative learning and group study habits.

Introductory Mathematics Courses

Numerous studies have shown that mathematics has proven to be a major barrier to increasing the number of (non-Asian) minorities and women in S.M.E. (science, mathematics, and engineering) majors. In this section we present information about some special programs, similar to Treisman's PDP program, that help women and minorities succeed in freshman mathematics courses. While there are more programs focused just on minorities, the University of Texas ESP program mentioned below and many other programs modeled on ESP target minorities and women as well as students from rural or inner-city schools.

Many of the new pedagogical approaches to teaching calculus and other introductory mathematics, such as cooperative learning, extensive writing, and open-ended projects, have been shown, in the words of Sheila Tobias, to "disproportionately benefit" groups underrepresented in mathematics. The study "Talk about Leaving", which is summarized in Chapter 22, states that "... switchers and non-switchers [out of S.M.E.] were almost unanimous in their view that no set of problems in S.M.E. majors was more in need of urgent, radical improvement than faculty pedagogy." In sum, efforts to help women and minorities succeed in mathematics are closely connected to efforts to provide high-quality mathematics instruction and to attract more students to enjoy and effectively learn mathematics.

The University of Texas Emerging Scholars Program

The following is a summary of the Texas Emerging Scholars Program (ESP). It is discussed in considerable detail in Chapter 12. The heart of ESP consists of intensive workshop sections that are supplementary to standard calculus courses. Each ESP section is led by an advanced graduate student with the assistance of two undergraduate ESP alumni. The main activity of a section is for students to work in small groups of 4 to 6 solving specially designed, challenging problems. These learning communities help ease the transition for the students from high school to a large university in all their studies. The section staff are available for support, and they help guide group discussions, but the students learn to be self-reliant in solving the problems.

The results have been impressive: ESP students have typically earned grades one half to one full grade point higher than the average of their mathematics class. Dropout rates for students in the ESP program are very low. ESP serves about 120 students at a cost of about $120,000 per year. While ESP strongly encourages participation of women and underrepresented minority groups, it is open to all students who meet the academic selection criteria and are willing to do the extra work.

The ESP program has been copied successfully at scores of institutions. Texas runs a training program on leading ESP workshops for faculty and graduate students from other institutions. For example, the University of Kentucky's MathExcel Program uses Texas's structure of three 2-hour workshops per week (instead of the regular two 1-hour recitations) and targets the same set of students. MathExcel continues into the second year with special sections of Calculus III and Calculus IV, but these do not entail extra class time. MathExcel workshop leaders are sent to the Texas training program. There are about 60 students in the Kentucky program, and the cost of the extra TAs and their training is about $33,000 a year, or about $500 per student per year. The results of Kentucky's MathExcel program have mirrored the successes at Texas in higher grades and improved retention. Since a number of other universities have also had very positive results with programs modeled on ESP, all mathematics departments should give serious consideration to starting similar programs.

The UCLA Mathematics and Science Scholars Program

The UCLA Mathematics and Science Scholars Program, abbreviated MS^2, was begun in 1992 by Professors Phil Curtis and Mark Green and Undergraduate Advisor Linda Johnson. It is a two-year intensive honors program that stresses academic excellence and professional development in mathematics, physics and chemistry. It was modeled on some other programs, especially the Uri Treisman program at Berkeley and the Texas ESP Program. The MS^2 program has three components:

PRISM (Pre-instruction in Science and Mathematics) is a two-week bridge program that takes place in the summer before the first quarter at UCLA. The students attend daily lectures in mathematics and chemistry and take field trips to nearby industrial sites to see how science is used in the "real world". PRISM students get free room and board on campus and an opportunity to meet other students with common intellectual interests before the school year begins.

EXCEL workshops (Excellence through Collaboration for Efficient Learners) are mathematics, physics, and chemistry workshops for freshmen and sophomores. Excel workshops are very similar to the ESP workshops, except that UCLA students receive no additional academic credit for the workshops and attend an ordinary one hour TA section or laboratory for each course as well as the four-hour-per-week workshop.

Mathematics 98 is a four-unit honors course taken during the first quarter at UCLA. Here students engage in research projects with a faculty sponsor and have weekly seminars by faculty explaining current research in science or by guest speakers who introduce students to the various careers available for mathematics and science majors. In addition, MS^2 students receive intensive academic counseling by the MS^2 director, priority enrollment in classes, individual tutoring, and a sense of community begun in the PRISM summer program and maintained throughout the year with several small social events.

Each year the program admits 50 freshmen (a total of 100 over two years) who are carefully selected by the director and a departmental committee for their ability, motivation and interest in mathematics and science. Through 1997 MS^2 admitted only minority or low-income students, but since 1998 the program has been open to all. MS^2 costs $1,000 per student, or $100,000 per year. It is funded by the chancellor, the dean of physical sciences, and the minority-oriented Academic Advancement Program of the Letters and Science College, and the Department of Mathematics ($30,000 of the total). The cost breakdown is PRISM $23,000, EXCEL $59,000, and Director $18,000. The director, a former UCLA mathematics major, devotes half her time to MS^2 and half her time as an undergraduate advisor in mathematics.

The MS^2 program is very successful in mathematics: MS^2 students who actively attend the Excel workshops get much higher grades, generally one half to one full grade point above the class. Curiously, the workshops have little effect on grades in chemistry or physics.

Mathematics Majors

Further along the pipeline, special attention is needed to increase the number of women and minorities majoring in S.M.E. disciplines and continuing on for graduate study. The successful mathematics majors, which are seen by students as welcoming and inclusive, have higher percentages of women and minorities than the typical mathematics major at a doctoral mathematics department (e.g., at Stony Brook, 18 percent are non-Asian minorities). UCLA's MS^2 program is not intended to recruit majors, just help students in lower-division mathematics and science courses, but over 60 percent of its students major in mathematics or science. The result of MS^2 combined with other strengths of UCLA is that 25 percent of UCLA mathematics majors are (non-Asian) minority students.

A successful program for minority mathematics majors at UC-Davis is described below. Many universities have some type of "Women in Science and Engineering" effort to attract more women into S.M.E. majors. We note that the University of Nebraska mathematics department's efforts to increase participation of women and minorities in mathematics were honored at the White House in fall 1998 with a Presidential Award for Excellence in Science, Mathematics, and Engineering Mentoring.

UC-Davis Minority Undergraduate Research Participation in the Physical and Mathematical Sciences (MURPPS).

MURPPS is a mentoring program designed to increase the number of women and minority students who obtain bachelor's degrees and pursue graduate work in the physical sciences and mathematics. It is sponsored jointly by the NSF, private foundations, and industry. It is directed by Emeritus Professor Henry Alder of the Davis mathematics department.

In MURPPS the emphasis is on individual research by a student paired one-on-one with an individual faculty mentor. Students receive a $600-per-quarter stipend, for which they are expected to spend ten hours a week working on a research project agreed upon by the student and the mentor and to participate in the Mathematical and Physical Sciences Seminar, which meets two hours a week for two quarters of the freshman year. In addition, MURPPS students may apply for a summer research internship, which pays $2,000 plus room and board. MURPPS participants present their research results every April at the UC-Davis Annual Undergraduate Research Conference.

MURPPS students are chosen from non-Asian minority and women students who are U.S. citizens or permanent residents. The students are expected to maintain a grade point average consistent with the admission requirements for graduate school and to have a satisfactory performance evaluation from their research mentor. In fall 1998 MURPPS had 29 students, including 10 freshmen.

Graduate Study in Mathematics

Finally, those minority and female students who enter graduate school in mathematics face major hurdles; too many leave without a Ph.D. There are often too few of them, there are few role models among the faculty to turn to for support, and, especially for minority students, their preparation is often below average. Along with having an overall friendly atmosphere in the mathematics department, most successful efforts for women and minorities at doctoral mathematics departments involve a committed faculty member, often a minority member. For example, Richard Tapia's great success in mentoring Mexican-American and Hispanic-American doctoral students at Rice has been recognized with a Presidential Award for Excellence in Science, Mathematics, and Engineering Mentoring (he was also featured in a PBS program two years ago). Nancy Kopell has been very successful in mentoring female graduate students at Boston University, 35 percent of whose graduate students are women. The success of such programs breeds future success, as prospective graduate students from underrepresented groups who visit the department see a welcoming environment with a critical mass of women or minority students.

University of Maryland

In trying to understand the causes for the high dropout rate among minority students at the University of Maryland, Raymond Johnson talked to faculty at Historically Black institutions. He identified another factor: that in graduate school the Black students missed the support network that other students had learned to build up as undergraduates. Johnson took it upon himself to sustain these students until they could build these networks, and organized small informal gatherings so that entering students could meet each other and more advanced students. (It should also be noted that the University of Maryland grants financial aid to students who display potential but who need to take some undergraduate courses in their first year.) The result has been striking. Of the twenty African-American graduate students with whom Johnson has worked, two have their Ph.D. degree, five have been admitted to candidacy, and five more are expected to pass their candidacy exams within one year. Another three or four have master's degrees, and some of these are pursuing doctoral degrees in other disciplines.

NSF's new Graduate Minority Education (GME) initiative for minority students in S.M.E. disciplines provides an opportunity for a doctoral mathematics department to collaborate with other S.M.E. departments at their institution for federal assistance in starting new programs to attract and retain more minority graduate students.

Broadening Graduate Education and Professional Development

One of the concerns voiced in several national studies of U.S. graduate education in the sciences and mathematics is that while doctoral students are excellently prepared to undertake research, they have minimal formal training in teaching. However, the vast majority of faculty positions are at colleges and comprehensive universities where quality instruction is the highest priority in the institution's mission. The Pew Foundation initiative, Preparing Future Faculty (PFF), has supported efforts to better prepare future professors for teaching and other aspects of their professional life in typical collegiate departments. To convey the importance and attraction of a rich professional life beyond the confines of the traditional research-intensive department, the University of Washington PFF program was linked with Seattle University, a medium-sized private institution, and Seattle Central Community College, nationally known for its innovative mathematics programs. As one of the PFF activities, selected UW graduate students are mentored for a quarter by individual faculty at a partner institution, typically meeting with mentors at least once a week on a project mutually agreed to. The Cornell PFF program is discussed in the following vignette.

Cornell University

The Preparing Future Faculty program in mathematics at Cornell has four general components. The first is an outreach effort to liberal art colleges. The primary vehicle is a program in which graduate students present lively mathematical talks to mathematics clubs and faculty at nearby colleges on, for example, topology and DNA. Graduate students also have informal discussions with students about what it is like to be a graduate student in mathematics. In return, during conversations with the faculty, say, over lunch, the graduate students learn about the life in a mathematics department in a liberal arts college. Another aspect of this outreach has graduate students going to regional meetings of the Mathematical Association of America to give talks to collegiate audiences, speak with undergraduates considering graduate study in mathematics, and interact with college mathematics faculty.

The second component brings college faculty to Cornell for job fairs. At the fairs, these faculty discuss the job market, requirements for tenure at their institutions, preparing CV's and cover letters, what they look for in letters of recommendation, teaching portfolios, and the like.

The third component involves instructional innovation. One year, faculty from Ithaca College presented talks about their NSF-funded calculus reform effort that involved student projects. The following year a group of Cornell graduate students took the initiative, with encouragement of Cornell faculty, to develop and teach their version of a reform calculus course with projects.

The fourth component involves a course in college teaching, which covers topics such as constructing course syllabi, cheating, handling obnoxious students, alternatives to lecturing, peer review, student mentoring, and the organizational structures of universities and colleges.

There has been growing interest in graduate training for something other than original research in mathematics. With the increasing use of quantitative methods on Wall Street, a few mathematics departments have started a financial mathematics master's program (a larger number of financial engineering M.S. programs have been started in engineering schools). Not surprisingly, NYU, two miles from Wall Street, recently started such a program. In this two-year M.S. program, students with a background in undergraduate mathematics take courses and seminars taught by Courant faculty and by investment bankers from the area. Those with a particular interest in computation may enter the scientific computing program, with a specialization in computational finance. The financial mathematics program at the University of Chicago is discussed in Chapter 10.

Industrial mathematics is another area of growing interest. The first and only Ph.D. program in this area is at the University of Minnesota. It was developed by Avner Friedman, director for the Institute for Mathematics and its Applications.

University of Minnesota

The Minnesota School of Mathematics offers M.S. and Ph.D. degrees in an industrial mathematics program. These are coordinated by the Minnesota Center for Industrial Mathematics, which is a center within the School of Mathematics. Ph.D. students in the program do a yearlong internship at an industrial research laboratory. M.S. students do a three-month summer internship. During the internship students develop a research topic leading to a Ph.D. or master's thesis. The MCIM's contacts with industry laboratories enables it to find suitable internship projects, chosen for their mathematical content. Participating companies include 3M, Bellcore, Honeywell, LORAM, Lucent, Ford, GM, Motorola, Deluxe Corporation, Lockheed-Martin, Computing Devices International, Medtronic, and Schlumberger. The students in the program have both a faculty advisor and an industry mentor. A university engineering laboratory can sometimes be substituted for an industry laboratory. Students' progress is closely monitored, and career development advice is provided. Graduates with master's degrees are often employed in industry, and Ph.D. graduates are well equipped for employment in both industry and academia. For more information go to

http://www.math.umn.edu/grad/.

Interdisciplinary Education and Research

The NSF initiative, Mathematical Sciences and Their Applications throughout the Curriculum, has given large grants to seven consortia to promote significant improvements in undergraduate education leading to increased student appreciation of and ability to use mathematics. NSF hopes that these projects will be models for better integrating mathematics into other disciplines, as well as for improving instruction in the mathematical sciences by incorporating other disciplinary perspectives. New courses, modules, software, and electronic materials are being developed by consortia centered at Dartmouth, Indiana, Nebraska/Oklahoma State, Pennsylvania, and West Point. The RPI consortium, Project Links, is developing a library of interactive multimedia learning modules integrating mathematical concepts with applications in science and engineering. The Stony Brook consortium is trying to achieve a broad, systemic change in quantitative instruction through diverse grassroots efforts by hundreds of faculty.

It is common in the sciences and engineering to see majors and graduate programs in new interdisciplinary fields develop, such as bioengineering and behavioral neurobiology. It is less common in the mathematical sciences. It is, however, common to see dual majors involving mathematics. Brown University makes a specialty of such majors, while the University of Washington provides an example of a truly interdisciplinary mathematical sciences major.

Brown University

Brown consciously promotes interdisciplinary education and research as mechanisms for effective utilization of small departments in a relatively small university. There are virtually no barriers to interdisciplinary and interdepartmental activities, and there is much encouragement.

At the undergraduate level, both the Department of Mathematics and the Division of Applied Mathematics offer a wide variety of interdepartmental concentrations (majors). Current standard concentrations include: mathematics-computer science, mathematics-economics, mathematics-physics, applied math-biology, applied math-computer science, applied math-economics (A.B. or Sc.B.), and applied math-psychology. The list of offerings has evolved over the last thirty years since Brown adopted its so-called "new curriculum". Obsolete programs go away, and occasionally a new program is started. Usually the catalyst for proposing a new interdepartmental program is student interest voiced in the new program. Some of these programs are small and geared to preparation for advanced study. Others are large (mathematics-economics and applied math-economics) and have gained a reputation as excellent preparation for careers in business.

University of Washington

At the University of Washington, the Departments of Applied Mathematics, Mathematics, Statistics, and Computer Science recently worked together to create a new interdisciplinary undergraduate degree program. The result is called the Applied and Computational Mathematical Sciences (ACMS) degree program.

The aim of the ACMS program is to provide a solid foundation in both applied and computational mathematical science with areas of application. A core set of courses in the basic tools common to many disciplines is followed by a broad set of pathways to suit different interests, such as statistics or mathematical biology. Flexibility in the requirements allows students with specific interests in another area to pursue a double major, such as ACMS and economics. The program seeks to prepare its students to pursue a variety of positions in industry after graduation or to go on to graduate or professional school in many fields. The

ACMS program builds on the strengths of existing departments and programs while presenting new opportunities for students. The interdepartmental aspect of the ACMS program stresses the unity of the mathematical sciences and provides a balanced education with a firm foundation in all aspects of applied and computational mathematics while also encouraging an in-depth study in some particular direction.

This new interdisciplinary major replaces the most popular current major within the mathematics department. Although it should attract additional students, one consequence may well be a decline in nominal total count of majors within the department. But the department felt that the value of ACMS and the connections with other departments far outweighed concerns about this decline. This cooperative attitude has already borne fruit. The participating departments used the ACMS program as the foundation for a joint proposal to the NSF VIGRE program. The "horizontal" as well as "vertical" integration of undergraduate, graduate, and postdoctoral training using ACMS as a core organizing principle was one very attractive aspect of the proposal, and it was one of six funded in the first round.

It is common for applied mathematics groups, either within a mathematics department or as a separate department, to have some interdisciplinary collaborations. Along with traditional applications of mathematics in the physical sciences, many new opportunities are developing in the biomedical sciences. NYU's Courant Institute is a leader in this area. Although Charles Peskin's models of the heart have garnered much publicity (and a MacArthur Fellowship), there are many other Courant faculty with biomedical interests, including one professor with a joint appointment with biology. The University of Southern California has a major multidisciplinary Center for Computational and Structural Genetics led by mathematician Alan Waterman. At Brown a research/graduate training initiative in applied mathematics, computer science, cognitive sciences and neuroscience has been awarded significant funding for graduate student support through the IGERT program at NSF and through the Burroughs Wellcome Fund. The cooperation among Ph.D. programs in this area is truly interdisciplinary, providing students in biological neuroscience, for example, with much broader preparation in mathematics and providing students in applied mathematics with opportunities for research on problems of theoretical neuroscience or cognitive processes (e.g., vision and speech understanding). Another collaboration among mathematicians, physicists, and biological/medical scientists at the University of Arizona led to the Biology, Mathematics and Physics Initiative, an IGERT-funded graduate program providing training opportunities in areas such as biomedical engineering, ecology and evolutionary biology, molecular biology, radiology, and neuroscience (funding was also supplied by the Finn Foundation).

Large multidisciplinary centers are starting to involve mathematics. Perhaps the best example is the NSF Science and Technology Center at Rutgers, known as DIMACS, that specializes in discrete mathematics and computer algorithms. The University of Arizona's Program in Applied Mathematics, building on Arizona's strength in optical sciences, formed the Arizona Center for Mathematical

Sciences, funded by an AFOSR University Research Initiative grant and other agencies. The center provides a research and graduate training environment in nonlinear optics, laser physics, and other nonlinear phenomena.

Collaboration with industry is an area of growing interest. The two groups frequently cited as leaders in this effort are the University of Minnesota's Institute for Mathematical Sciences and Stony Brook's Department of Applied Mathematics and Statistics. Stony Brook has had about thirty industrial collaborations in the past few years, many with companies located at the other end of the country.

Putting It All Together

This section presents two vignettes of mathematics departments that have succeeded in bringing together a number of the different components described in this book; they are viewed as a model department by their university administrations and other departments. One, the Nebraska Department of Mathematics and Statistics, has had a reputation for being a successful department for several years. The other, the Rochester mathematics department, has gained this status at its institution quite recently. Happily, there are many mathematics departments that have been successful in achieving this goal. For example, most of the departments mentioned in this chapter and the departments that the Task Force visited are viewed as quite successful by their administrations.

In some cases this success is closely associated with a highly effective chair. However, the reality is that successful chairs cannot exist without the strong support of senior faculty and an activist climate collectively generated by all faculty. Conversely, while the chairs of many successful departments are not well known outside their campus, they have almost all become very effective in dealing with their administrations and other departments. They make sure that their departments respond constructively to the concerns of administrators and other departments at the same time that they communicate the achievements of their departments. These skills are difficult to master in a three-year term as chair.

University of Nebraska-Lincoln

In the late 1980s the Department of Mathematics and Statistics at the University of Nebraska-Lincoln developed a strategy of using success with undergraduate instruction and outreach activities as the means to secure resources from the University that might also benefit research and graduate education and move the department up to Group II status in the next NRC rankings. The department's stated goal was "to become a model department of mathematics in a research university where educational goals are integral to the departmental mission and are supported by broadly based participation in educational programs."

The most important change in the department has been a change in the culture. While the department chair (who is a member of this Task Force) enjoys a reputation as an effective chair, he is the first to give credit for the change to the faculty as a whole. Most faculty now have a strong commitment to being an excellent teacher in addition to a commitment to developing a high-quality research program. Moreover, many faculty are involved in at least one educational program that extends beyond a basic commitment to being a good teacher. Half the faculty have won a College or University Distinguished Teaching Award. In 1998 the department won the University of Nebraska's University-wide Department Teaching Award.

The department has rethought its instruction at all levels. Its below-calculus offerings have been restructured, with failure rates cut in half. The graduate program also benefited because the UNL administration provided $90,000 in additional TA support for this initiative. Mathematics faculty played an important role in advocating an increase in UNL's admission requirements, which substantially reduced the percentage of students placing below college algebra. In response to a campus-wide general education initiative, the department introduced a contemporary mathematics course (based on the text *For All Practical Purposes*) to meet the needs of students in the humanities, arts, and education. Because of the substantial demand for this course, the administration provided (over time) funds to support three new GTAs, one postdoctoral position, and one new tenure-track line.

The department reworked calculus with funding from the dean, provost and UN Foundation. The effort was led by a group of younger faculty, all with NSF research grants. As well as introducing a reform text and graphing calculators, they sought to create as active a learning environment as is possible with a large-lecture/recitation format. There are small-group work and extended writing projects which help calculus meet the expectations of the general education initiative. Another faculty member developed a Web-based system for giving "gateway exams" in calculus to test technical skills. Students are allowed to retake the exam many times until they meet the department's high standard. Ten other UNL departments are using the gateway software, and John Wiley is marketing the UNL calculus gateway exams nationally. On one recent day, the gateway server had 36,000 hits! An NSF grant supported the introduction of computer algebra software in differential equations and matrix theory courses. To help mathematics instruction connect better with other disciplines, the department secured a large grant, jointly with Oklahoma State, in the NSF Mathematics Across the Curriculum program to develop interdisciplinary courses.

An important component of the department's contribution to the University has been its impressive array of outreach activities. Most famous is the American Mathematics Competitions, which Walter Mientka ran at UNL for three decades. With special funding from the state legislature, the department initiated in 1989 a mathematics prognosis test for high school juniors, modeled after programs at Ohio State and LSU. Its UNL Math Day brings approximately 1,300 students to campus each year to compete in mathematics competitions. The program receives outstanding cooperation from the University because of its potential to help recruit outstanding students. Recently, UNL Math Day became sponsored by The Gallup Organization.

The most substantial recent outreach program was the Nebraska Math and Science Initiative. The department received $10,000,000 in NSF funding to establish a statewide systemic initiative involving school mathematics teachers across the state. Over time it grew to include a science component. One NMSI effort, a middle school videotape curriculum project called Math Vantage, won several national awards. The videotape is now producing royalties for an endowment to support math education activities in the department. After NMSI's NSF funding ended, the chancellor reallocated $150,000 (annually) to provide a permanent infrastructure for the NMSI Center, and the dean of Arts and Sciences designated math/science education an "Area of Strength", with an annual budget of $70,000 for the math and science chairs to support educational activities.

These and other educational activities have improved the instructional environment for faculty, helped attract better students, raised the level of external funding, and secured additional graduate TAships and faculty positions, while greatly enhancing the department's standing on campus. At the same time, the department recommitted itself to improving its graduate education and research productivity. Along with producing more Ph.D.'s, the department made a priority of recruiting and encouraging female graduate students. While in the 1980s the department produced only 23 Ph.D.'s, none of whom were women, it has produced 34 Ph.D.'s in just the past four years, 13 of whom were women, one of whom won an NSF postdoctoral fellowship. Currently half of the graduate students are women. In fall 1998, the department won an NSF Presidential Award for Excellence in Science, Mathematics, and Engineering Mentoring in recognition of its success with women.

In the past three years 75 percent of the faculty have had some type of external funding (including education grants). The department has taken the lead in a College-designated Area of Strength: discrete, experimental, and applied mathematics. Other faculty are actively involved in initiatives like the Gallup Research Center. The A&S College has been very supportive of the department's efforts to enhance its research standing—for example, by allowing it to make competitive offers with good start-up packages. The 1995 NRC ratings placed the department solidly in Group II.

University of Rochester

In 1998 the mathematics department at the University of Rochester won a recently established $30,000 University Award for Curricular Achievement. While it was under severe attack from its administration in 1995, it is now seen by them as a model department. The criticism of the department, which focused on calculus instruction and isolation from other departments, had the effect of uniting the mathematics faculty to undertake a multipronged effort to change its image on campus. Fortunately, several major initiatives, including WeBWorK (described below), were under development at the time and enhanced the impact of the department's renewed dedication to providing high-quality mathematics instruction.

At the time of the attack on the department, the University announced its Renaissance Plan to cut the incoming class size by 20 percent while simultaneously aiming to attract better students. The department aligned its instructional mission with this goal of educating stronger undergraduates. This plan has served it well, since mathematics enrollments have increased despite the smaller number of students. The department has become the most active participant in the University's new Quest program of lower-division courses that bring research into the classroom. The provost at Rochester now touts the fact that 10 percent of the current freshman class (triple the historical average) are in honors calculus classes that include special research-like workshops. In addition, the dropout rate in these courses has been cut in half.

The department took steps to address the perception as well as the substance of calculus instruction. Meetings were held with every department having a math requirement for its major. Two new courses arose from these discussions, along with a number of personal connections that have been maintained to coordinate mathematics instruction with courses in other subjects.

They also entered into a constructive dialog with a vocal critic of mathematics instruction who has a degree in applied mathematics and was teaching competing courses in the mechanical engineering department. This led to a joint appointment for him in mathematics, and mechanical engineering is now phasing out its mathematics offerings.

Among the many tangible results of the improved communication and documentation of department efforts are the curriculum award cited above and two $5,000 University teaching awards (five of which are given out each year) for mathematics professors in the past two years. The department was also successful in nominating one of its faculty for the MAA Seaway Section Award for Distinguished Teaching in 1997.

One of the criticisms of calculus instruction at Rochester was that homework was not being graded. The department has come up with a remarkable solution to this problem. Mike Gage, a faculty member who earlier in his career had worked as a systems programmer for Intel, led the development of a complex software package called WeBWorK that allows students to do their homework and have it graded through the Internet. Parameters can be randomized (within set limits) to give each student his or her own individual set of weekly homework problems. A student who enters a wrong answer can try again any number of times up to a deadline date. This immediate feedback has proved to be a strong motivational tool. Currently, 75 percent of all students in freshman calculus are working on their WeBWorK homework assignments until they get every problem right.

The Rochester Physics Department now uses WeBWorK, and several mathematics departments at other institutions have started using it. It has also sparked interest in some high schools. One high school teacher wrote, "The WeBWorK project is easily the most exciting and potentially beneficial use of web-based technology for education I have come across." Interested readers are invited to contact Arnold Pizer (apizer@math.rochester.edu) or Mike Gage (gage@math.rochester.edu), or to browse

http://www.math.rochester.edu/WeBWorK/

using "practice1" as a login name and password.

Part IV

Views

Chapter 14
How Do Departments Survive

William Kirwan[1]

I am very pleased to have an opportunity to participate on this panel. One of the worst aspects of my present position is the degree to which I have been distanced from direct involvement in issues affecting mathematics.

The subject for the panel is obviously very timely. However, the title has a somewhat more optimistic tone than I believe is warranted. Perhaps I am a bit jaded from my experience with the budget cuts at College Park and at other universities, but for me a more fitting title might be "How Do Mathematics Departments Survive During a Time of Diminishing Resources and Declining Public Support?" Whatever the title, I believe the topic for today's discussion and the work of the AMS Task Force on Excellence are extremely important for the future well being of our discipline.

I would like to focus for a few moments on some of the resource-related issues that we as a community face now and probably will face for the rest of this decade. First, the obvious: we have needs and demands for expanded activities that far outstrip available resources. A recent survey conducted by the American Council on Education determined that 47 percent of public four-year colleges and universities have flat or declining budgets. I am confident that the data specifically for mathematics departments are no better. The stories of resource strain in universities from Maryland to California and from Oregon to Florida are well known, and the situation is not likely to improve in the near term. John Wiesenfeld, Cornell's vice president for planning, was recently quoted as saying, "We are looking at a sea change in the environment for higher education, both private and public. Understanding the implications of these changes," he says, "is now what we must do." So, the first issue the AMS committee must face is the apparent reality that, in terms of available resources, the 1990s are going to be far

[1] William Kirwan is currently the president of The Ohio State University, as well as past president of the University of Maryland. This essay is based on a talk given during a panel discussion sponsored by the AMS Committee on Science Policy at the Joint Mathematics Meetings in San Antonio, Texas, January 1993.

bleaker than anything post–World War II trained mathematicians have yet experienced.

Comparatively speaking, this is the good news in my observations. Let me now turn to a second challenge that we as a community and the committee face. Perhaps the only thing falling faster than our resource base is public understanding of and support for the work we do at research universities. Our situation is perhaps best summarized by the "hearings"— and I use that term advisedly— conducted by Congresswoman Pat Schroeder this past fall on the state of undergraduate education.

There are many advantages to being a research university in the Washington, DC, area. But especially when it comes to congressional hearings on sensitive higher-educational issues, there are also disadvantages. If you want to bash universities, where do you turn? Obviously to the editor of the school newspaper at one of the local research universities—in this case, College Park. Never mind that the testimony that this student and others gave was, at the very least, overstated. The fallout from this hearing and other hearings on the same subject now being planned by Congressman Dingle and likely to be emulated in state legislatures across the country could do considerable further damage to our image, an image that also has been tarnished by research fraud (not, I am proud to say, in mathematics) and by excesses of administrators in the use of research overhead. Charles Vest, president of MIT, said it well the other day in testimony before a White House panel: "Growing out of a sense of disappointment and mistrust, research universities rest on unstable and shifting ground."

The focus of much of the criticism of research universities is the lack of attention given to undergraduate education. For some this gets translated into faculty "teaching loads", so we see legislation in states like Florida, Oregon and, I believe, California, mandating increased teaching loads. In Maryland and in many other states, legislators are asking for information on teaching loads—note I said "teaching" and not "work" loads. Many of our critics do not understand the difference.

But it is not just aggressive and somewhat uninformed legislators who are critical of the quality of undergraduate education. Leaders of our most distinguished research universities also have spoken out on this topic.

For example, in a recent article in *Change* magazine, Derek Bok, president emeritus at Harvard, cited the lack of attention to undergraduate education, primarily at research universities, as the number one issue causing the decline in public trust of higher education. He said, "Until we convince the public, by our actions, that we indeed make education our top priority, that we are committed to the highest quality of undergraduate education, we will continue to be vulnerable to attacks on our curricula, our faculty, our tuition, and all the different issues on which we have been taking punishment the last few years."

Richard Atkinson, a member of the National Academy of Sciences, president of the University of California, and former director of the National Science Foundation, said something similar in an article he published with Donald Tuzin of UCSD. They wrote, "...research universities should lead the way by restoring the balance between teaching and [research]." They go on to say, "...the continued greatness of the American research university depends on ... an equilibrium between the three missions of its charter—the propagation, creation and application of knowledge. When the balance goes awry, the entire edifice erodes. The chances of collapse may be slight, but the dysphoria has gone on long enough. It is time to re-establish equilibrium."

Much the same view is expressed in the just released report of the Presidential Commission on the Advancement of Science and Technology, a commission established under the leadership of science advisor Alan Bromley. The report, entitled "Renewing the Promise: Research-Intensive Universities and the Nation", makes an eloquent case for the role that the nation's research universities have played in the advancement of our society. The report also addresses the issue of instruction at research universities. Among many recommendations, it says that universities must:

- increase direct senior faculty involvement at both the undergraduate and graduate levels and in counseling students;
- balance the contributions of teaching and interaction with students with those of research and service in evaluating and rewarding faculty;
- place less instructional emphasis on graduate teaching assistants;
- develop new pedagogics for undergraduate teaching;
- assist with national, state, and local efforts to revitalize precollege education in science and mathematics; and
- provide incentives for outstanding undergraduate and graduate teachers.

Thus, it seems clear that the AMS Task Force on Excellence must deal substantively with the issue of the quality of undergraduate mathematics education for all students, not just mathematics majors. The general population, who in the final analysis is our source of financial support, is demanding that this happen, and many of our most respected academic leaders concur. We can resist these demands, but, in my view, we do so at great risk to our discipline. Unresponsiveness on our part and further alienation of the general population toward our research universities is likely to lead to even more onerous externally imposed "workload" requirements and further declines in our support base.

There is a third issue, related to the previous one, that I believe the AMS Task Force must consider. This is the role of mathematics departments in reform of K–12 education. This is yet another demand being pressed upon us which, in my view, we cannot avoid. There is a very definite movement sweeping the nation calling for the elimination of the bachelor's degree in education. To a large extent, this movement has been spawned by a group of our nation's best colleges of education. This group, known as the Holmes Group, now numbers more than one hundred. A central principle of the group is that K–12 teachers should get their first degree in an academic department with support work in education. The State Board of Education in Maryland and boards in several other states are presently considering proposals to modify teacher certification along these lines.

Based on what I have heard in Maryland, there is significant public support for such a reform. If carried out, this change would of necessity bring mathematics departments into a much closer working relation with the K–12 sector. To be sure, there are already significant school/university initiatives at many of our research universities. But the change I foresee could lead to substantially increased expectations for our already overburdened mathematics departments. Of course, these expectations also create considerable opportunity for research universities, especially in mathematics because of its central role in education at all levels. There is a chance we can effect positive change and increase public awareness of and appreciation for our discipline and our institutions.

Even if the view of the future I have described is only partially correct, it seems clear that the AMS Task Force has a formidable challenge: providing recommendations in an environment where there will be fewer resources in absolute terms and greater demands on our departments.

How are we as a community to cope with this situation and maintain, as we must, the vitality and evolution of our discipline? Despite the generally bleak picture on resources, there should be incremental funds available for improvements to undergraduate education. First, the science education division is one of the few divisions in NSF with a hefty budget increase. Also, university administrators are under considerable external pressure to demonstrate commitment to undergraduate education. Since failure rates and attrition tend to be high in lower-division science and mathematics courses, proposals to improve the quality of these courses are likely to receive a favorable response. For example, I believe that Indiana University recently invested significant new resources for reforms in the calculus sequence. Of course, it probably helped that the dean of the college is Mort Lowengrub, a mathematician.

Another alternative that deserves consideration is to modify the reward structure at research universities for tenured faculty as a means of encouraging some faculty to devote most of their energies to teaching and curricular matters. Obviously, such a move is an issue for individual institutions and departments to decide. And it is my understanding that several universities are beginning to explore proposals in this direction. In mentioning this idea, I emphasize tenured faculty because I believe a research university must insist that those to whom it grants tenure demonstrate a mastery of some important subdiscipline of their fields.

Do not misunderstand what I am saying. We should continue support for the most talented researchers, especially the youngest of these individuals, more or less as we do at present. But, in my view, we must make it easier for senior mathematicians at research universities to take on with dignity, respect, and reward some of the challenging obligations facing the mathematics community. I believe there is food for thought in a recent address by Don Kennedy, former president of Stanford. He said that "the overproduction of routine scholarship is one of the most egregious aspects of contemporary academic life: it tends to conceal really important work by its shear volume, it wastes time and valuable resources, and it is a major contributor to the inflation of academic library costs." In the article by Atkinson and Tuzin that I cited earlier, the authors say a similar thing: "Research universities can relieve the strain on resources by honing the research enterprise to redirect the work of individuals whose energies could be better spent in other areas of the university's mission."

I believe there is an especially important role in this general area for the AMS. As the primary professional society for research mathematicians, the AMS is in a position to exert great influence on the community. The Society's support for an expanded reward structure and its recognition of exceptional contributions to mathematics pedagogy would go a long way toward creating an environment where the changes I describe can occur.

There is one final point I would like to make. In my view, as a group mathematicians have done a poor job of explaining to the university community and the general public the value of the work we do. The AMS needs to consider ways in which our, community can better articulate proactively the value we add to the intellectual base of our nation. I fear that we are losing out in the struggle for support between the advocates for "big science" on the one hand and, on the other hand, the proponents of research expenditures tied more closely to the nation's economic growth. We need to make a better case for the intrinsic value of mathematics and, in particular, mathematics research.

In conclusion, let me say that in comparison to other disciplines I believe the mathematics community has demonstrated a remarkable degree of responsibility and leadership in its willingness to address the difficult issues facing higher education. Among other efforts, the "David Report" and MS 2000 reports, the development of the new NCTM standards, the appointment of the AMS Task Force on Excellence, and the JPBM Committee on Reward Structures are indicative of an academic community responsibly grappling with its future in these uncertain times. These and other efforts make me feel proud to be a mathematician.

Chapter 15
A View from Above: Interactions with the University Administration

Ettore F. Infante[1]

A department of mathematics, as its name implies, is a component of the university of which it is a part. Whereas the mathematical profession transcends institutional—indeed, national—boundaries with its research activities, values, culture, rewards, and means of interaction, the department is local, embedded, and largely dependent on the university of which it is a part for resources and infrastructure support. For a department to be successful it must be able to manage—indeed, to appropriately leverage on each other—the expectations, values, rewards, and resources of the university of which it is an integral part with those of the larger disciplinary world to which its faculty belongs. This is a particularly critical task at research universities, with their dual mission of research, which transcends the particular university, and education, which is more local. It is thus important for a department and its leadership to develop a clear understanding of its institutional setting, of the stated mission of the university of which it is a part, and of its role within it. Effective communication within the administrative structure of the university depends on it.

This brief presentation of a "view from above"—that is, of the context and criteria with which deans, provosts, and senior university administrators interact, view, evaluate, and prioritize resource allocations to a department and its activities—is intended to help faculty and departmental chairs better understand this process.

A useful maquette that captures the essence of the context within which university administrators view a department can be expressed by three words: mission, money, and impact. These words refer to three highly interrelated aspects of a university. The mission of the institution is the basic compact between it and the larger society that provides it with resources and support, and the term "impact" includes the quality, effectiveness, and efficiency with which that mission is discharged through the use of the financial resources that are provided. Deans,

---

[1] Ettore Infante is currently professor of mathematics and dean of the College of Arts and Sciences at Vanderbilt University.

provosts, and senior university administrators have the tasks of being the conscience of the institution as to its mission and use of resources and of acting as spokespersons for the university to external constituencies as to its values, accomplishments, and resource needs. Perforce, deans and senior administrators are at the center of resource allocation to depart,ments and are the ones most accountable for these resources. Within this context it is their task to see that clear answers are provided to the litany of questions—who pays? who benefits? who should be subsidized? from what sources? for what purpose? and with what impact?—and to see to it that these answers are sustained by the enduring values, ideas, and ideals of education and scholarship.

The mission of the university, intimately tied to the sources of funds that support it, is central to any communication between departments and academic administrators. At research universities the mission is multifaceted, concerning discovery and learning, dissemination and teaching, and the promotion and use of knowledge in society. With notable exceptions, research universities have mission goals in undergraduate education, including general education; in graduate training; and of course in research and outreach. It is essential that faculty and departmental leadership have a clear understanding at their university of institutional expectations within these components of the mission, of the sources of funds that support them, and of the role that is expected from the department. Discussion of mission, roles, and responsibilities is the essential base for appropriate interactions between the department and university administrators. On this base, further interactions center on impact and on the resources needed for appropriate impact.

It is useful to differentiate five aspects of impact: centrality, quality, effectiveness and efficiency, demand, and comparative advantage.

Academic administrators must of necessity pay particular attention to those university activities and structures that rank high in centrality. Mathematics, as a discipline, shares the distinction of a high level of centrality with English and the library, for it plays a very particular role in general education and an essential role in the preparation of a large spectrum of students for further study in the sciences and engineering. This centrality of mathematics results in concerns, expectations, and willingness to invest by deans and provosts that go beyond those directed to units with less widespread academic impact on the entire institution. It also leads to the fact that a dean cannot but have a high level of concern for the performance of her mathematics department in undergraduate education, for it must be noted that the centrality of mathematics is most evident in undergraduate education, much less so in graduate education and research.

Quality, in the eyes of administrators, is the result of an evaluation of the outcomes of the activities of the department. The external reputation of the departmental research activities and of their impact on the national and international research and applications community, the quality of the preparation of undergraduate and graduate students produced as reflected by their professional contributions and accomplishments after graduation, the satisfaction of other departments within the university in the mathematical preparation provided to their students, and the leadership of the department in outreach activities in education and in multidisciplinary research are some of the elements that underpin the judgment of the quality of a department. It is essential that department chairs provide appropriate information to their deans on which appropriate judgments can be made. Not only should information about achievements and successes be provided, but also credible, realistic appraisals of shortcomings together with plans to alleviate them. Impact of high-quality merits reward; its sustenance requires resources.

Quality plays a most important role as a criterion in the evaluation of plans and budgets. So do the criteria of effectiveness and efficiency. How effective is the department in its manifold tasks of undergraduate and graduate education, of research and outreach? What goals and strategies has the department set for student recruitment and retention, for rapid progress through their studies, for the utilization of technology and of innovative teaching methodologies, and for the securing of external support for research and educational activities? Efficiency refers to the cost-effective utilization of resources in the pursuit of goals by the department. Plans and budgets must represent the embodiment of considerations of quality, effectiveness, and efficiency. Discussions centered on planning and budgeting are an opportunity for a departmental chair to engage in meaningful communication with deans and senior administrators as to the assets and needs of the unit.

Finally, a dean, faced with the always difficult choices implied by resource allocations, will want to address issues of demand and of comparative advantage. What demand is there from students and other departments for the instructional services provided by the department? What is the demand for doctoral students, for research activities, and for outreach? Chairs should be prepared to document existing demands, as well as realistic opportunities for the department and the university to undertake new and novel activities in response to felt needs. Comparative advantage, as the term implies, is a judgment on the part of administrators that leads to preferential investments in a particular area or department because of the belief that some sort of benefit-cost ratio will be maximized through that investment. Comparative advantage is most often based on an evaluation of strength, seldom of weakness; on evidence that the department has well-laid plans which it is already implementing through the reallocation of its own existing resources, thus demonstrating high priority; and on how strongly the resources the dean is asked to invest will leverage other activities of high priority.

Planning and budgeting are yearly opportunities for a department to present its case for the centrality of its activities; for their quality, effectiveness and efficiency; and for its plans to respond to demands and opportunities based on its comparative advantages. Discussions between chairs, deans, and provosts on budgets are perforce based on data. University administrators are most knowledgeable about facts and data internal to the institution; much less so about data on mathematics departments at peer universities. It is the essential responsibility of the department and of its leadership to develop such cross-institutional data and to present it to university administrators. Credible "benchmarking" with peer departments on resources and performance should be developed to address the five criteria of impact previously described. Most helpful is the result of an external review of the department, where qualitative evaluations are based on benchmarking data developed as part of a departmental self-study. The most effective comparative data is centered on competitive situations: success in grant activity, in publication of research, in placement and success of graduates, in philanthropic fundraising, and in reputational rankings. But it is most important to provide comparative data on all resources and results. University administrators must make decisions on resource allocations within their own university; in so doing they are driven to comparisons and evaluations of the diverse disciplinary units for which they are responsible, yet they are committed to the competitiveness of these units with their peers at other institutions. It is the task of the chair of the department to provide the data and information so that deans and provosts can reach informed judgments; no one else but the chair, with appropriate help, can undertake this task. For mathematics, with its highly developed and somewhat unique role within the disciplines and the university, this is a crucially important task.

Matters of mission, money, and impact are at the heart of communication between chairs and academic administrators and are central to the evaluation and resource allocation process. There is another element that plays an unusually important role: the perception by administrators and department outsiders of the department's "atmosphere"; of the quality of the interactions within the unit and with other departments; and of the reliability, credibility, and stability of the senior faculty and the departmental leadership. Trust is the golden coin of the academic realm; civility and responsiveness to the needs of the institution are essential to its flow. Often departments have been judged as less than successful and not deserving of resources by being perceived as fractious, isolated from the rest of the institution, unable to set goals and priorities, and unwilling to be guided by long-term leadership. Deans are known to speak of the "culture" of departments, sometimes in negative terms, but also sometimes in admiring ones. A positive, responsive, and civil culture within a department and long-term responsible and foresighted leadership by the senior faculty and the chair are important to the success of a mathematics department within the modern research university.

This said, the five criteria described and the appropriateness of the role of the department within the mission of the university constitute the basic elements that underpin the discussions on the evaluation and resource allocation to the department by academic administrators. Successful discussions are essential to the well-being of the department.

Chapter 16
A View from Below

Doug Lind[1]

Recently I completed a five-year term as department chair, and I have been reflecting on what I would have liked someone to tell me before I started out. I picked up some of what follows at the BMS chairs colloquia, some from talking with other chairs, some from the Task Force focus groups, and some from bitter experience.

1. Work Very Hard at Your Relationship with Your Dean

The relationship between the chair and dean is crucial to the health and success of the department. They should agree, at least in general terms, on the mission and goals of the department and how to measure progress. In case after case, the ability of a department to change and prosper has depended on the dean trusting the chair and feeling that the department was accountable. Striking examples of this are Don Lewis, followed by Al Taylor at Michigan; and Bus Jaco, followed by Brian Conrey at Oklahoma State. The dean at Oklahoma State said he knew "how good the department was in keeping the wolves away from the door in terms of when we talk with the state people." Considering the resources Oklahoma State has to work with, they have done incredible things.

On the other hand, plenty of deans told us about their frustrations with mathematics departments, complaining about their insularity (one dean said that the department doesn't talk among themselves, much less with other departments), their not taking teaching basic courses seriously enough (as evidenced by widespread complaints from students and other departments), frictions between mathematics and applied mathematics (sometimes so disastrous as to cripple the department), and their nostalgia for the good old days which will never return.

It is essential to this relationship that chairs and deans understand each other's needs. It does no good for a chair to push for increased research support if the dean's main worry is precalculus instruction. Fitting the department's goals within the overall missions of the university first requires the department to un-

[1] Doug Lind is currently professor of mathematics at the University of Washington, where he has recently served as chair of the Department of Mathematics.

derstand what those missions are. This means involvement of faculty beyond the confines of the department (e.g., talking with other departments, the faculty senate, regents' meetings, even the legislature). A chair should meet regularly with the dean to discuss what they are doing and should modify this in light of any new information. This is a two-way street, and a chair should not be an administrative toady.

Deans like named things. So instead of proposing simply to improve precalculus instruction, call it Project PreCalc, with specific goals, faculty, budgets. This is something a dean can brag about to the central administration and other deans. It also serves as a focus for funding. Make the dean look good.

Deans also look for departments to prioritize and make choices. Ending one department activity in order to fund something more important tells a dean that the department is responsible and is willing to take a hard look at itself. Not every new project should involve major new money from the college (you won't get it). Outside funding can be crucial in getting a project off the ground (e.g., Texas Instruments money to start the Mathematics Learning Resource Center at Oklahoma State, later sustained by student fees).

Find out what data the dean is using to judge the department. Is this data shared with the department? How are comparisons with other departments made, and do the chair and dean agree these comparisons are fair? You should also compare notes with other chairs to check for consistency.

Keep the dean informed of potential trouble and how you're handling it (e.g., potential sexual harassment charges, uprisings by undergraduates, threats of lawsuits, etc.). The last thing you want is for your dean to be blindsided by a very unpleasant event.

2. Hone Your Negotiation Skills

Your success will largely be determined by your skills at negotiation. Buy and study *You Can Negotiate Anything!* by Herb Cohen. Understand that the three keys to negotiations are time, power, and information. Knowing how these work in a particular situation can go a long way towards success. Strive for "win-win" outcomes.

3. Understand the Position of Your Department in the University

Meet people from around campus informally (say lunch) to get to know each other. These could include engineering and education deans, a vice provost for undergraduate education, chairs of other departments, faculty, staff, students, and so on. This can be enormously interesting, and I found it one of the real joys of the job. It also makes it much easier to call someone later to ask a favor or get some key information and for them to do the same with you. One thing you should strive for is a frank expression of how others view the department. If this is favorable, it's good to know, and if unfavorable, it should set off alarm bells that demand attention. It was amazing to me how little departments know about each other and their very different cultures.

It also helps to know how power works in your university. Who controls what resources and how are they allocated? Involvement in the faculty senate or even attending regents' meetings can be quite enlightening.

4. Burnish Your Department's Image

Nominate your best faculty and students for awards, both internal and national. Tell them you're doing it. Set up your own awards for students and create a ceremony, which potential donors should be invited to. Be a PR person for mathematics within your community. Learn whom to contact in the local press with story ideas, and build relationships with them (for example, by using the occasion of a conference on campus, I once got a detailed story about the Riemann Hypothesis on the front page of the *Seattle Times*).

5. Data and Budget

Good data is a golden currency when making your case. Data that meshes with your administration's is even better. Know your department's budget, and get monthly statements so you have a sense of how it's being spent. Fundraising will be increasingly important to enable you to do those wonderful discretionary things that give you a warm glow inside. It's hard to do and a long-term effort, but your college should help. Find out who's been successful, and how.

6. Deal with Stress

Let's face it, chairing a large department of colleagues is a very tough job. You will have to make important decisions about the lives of the people you know and live with the fallout after you've stepped down. The Golden Rule is useful to remember: treat others as you would like to be treated. But the accumulating stress can cause all sorts of problems, and you should be aware of signs when things are getting bad and take steps to manage stress (running, sports, massage, whatever). You are no good to anyone if you're so wound up you can't think straight.

7. Take Pleasure in Making Your Department a Better Place

As chair you can play a huge role in making your department a better place for its faculty. In literally hundreds of ways you can bring out the best in your colleagues, providing support, encouragement, ideas, and sometimes constructive criticism. Take pleasure in this, for it will tide you over the rough spots.

Chapter 17
Communicating with the Administration

Alan Newell[1]

Without exception, successful departments have established credibility with the university administration and particularly with the levels of dean and academic vice president. They have done this by recognizing clearly their unique position (the centrality of mathematics) and the awesome responsibility that goes with it. They have not waited to be asked, coaxed, prodded, or coerced. Rather, they themselves have taken the leadership in addressing the enormous range of challenges bestowed on a department in a Research One University, responsible for the literacy, consciousness, and education of a generation of students of widely varying abilities and the propagation of knowledge both within the discipline itself and across disciplinary boundaries. It has often been said that mathematics is far too important a subject to leave to the mathematicians. The successful department gives lie to that statement by accepting the role as quarterback and by clearly defining goals, strategies and plans for meeting the expectations placed upon it by the overall mission of the university.

The other components of the university structure, from central administration to client and other disciplines, do not resent such precociousness. On the contrary, they welcome such initiative with open arms. We cannot overemphasize the enormous leverage a department can gain by establishing its credibility and competence in handling its mission. The palpable and collective sighs of relief coming from the carpeted corridors of power in central administration are clearly audible. They know that failure to provide an effective preparation in mathematics for its undergraduate population is guaranteed to give presidents, vice presidents, and deans endless hours of headaches generated by complaints from students, parents, and state legislators. And we know it too. And therefore we know that by relieving them of the burden of concern over undergraduate mathematics and by establishing a bond of trust that mathematicians can develop strategies to further the overall university mission, deans and academic vice presidents will be predisposed to listen sympathetically to well-argued and sensi-

[1] Alan Newell is currently professor of mathematics at the University of Warwick, England, where he also serves as chairman of the Mathematics Institute.

ble plans for resources to cover and advance the entire spectrum of departmental goals.

And the beauty of things is that this contract with central administration need not involve a whole lot of new resources. To be sure, we have found that successful departments have needed some additional monies to develop programs that provide more personal attention for students in entry level mathematics and to seed efforts to improve the computational environment. But the most important resource for the mathematics department in a Research One University is people, and the most important currency is positions, to renew the lifeblood of the department with new young faculty and to attract a regular stream of first-rate visitors. To provide these kinds of support requires almost NO NEW MONEY. All it requires is NERVE and a belief in statistics on the part of the dean and academic vice president. A little analysis of most major departments will show that if the university is willing to commit replacements and the use of funds generated by unpaid leaves of absence, then the dean can guarantee the department for the next N years that it can recruit n new tenure-track positions per year and m temporary visiting (with teaching responsibilities) positions. In the case of many of the departments we surveyed, N was five, n was at least two, and m at least four.

A contract which, in return for a clear and sensible departmental plan, guarantees a reliable stream of concrete funds has proven to be invaluable for a multitude of reasons. First, departments can plan ahead. They can actively seek out the best new talent and make concrete offers at any time. They can get the best visitors because they can make their offers early. Second, and most important, the knowledge that there is a stream of openings on line removes from departmental deliberations one of the main ingredients of dissension, namely, the belief that each appointment is the last and that different specializations within the discipline are doomed if they do not capture the positions for themselves. The certainty of positions means that each of the areas declared to be priorities can wait until it has found the very best person rather than push those less-than-perfect cases for reasons of territorial gain. Indeed, we have observed first-hand the presence of a spirit of cooperation in departments which have long-term strategies underpinned by real and concrete financial support. Third, and especially important, it allows a department to adopt genuine change, to make plans to test the waters in new areas. In particular, it helps a department build ties with other disciplines. In one case we know, a mathematics department was willing to use one of its positions to attract a couple of new people in financial mathematics. The other position was supplied by the business school. Each department will immediately gain twice the value of its involvement. Fourth, it enables the department to foster links within the subject itself which promote and celebrate the unity of mathematics. Anachronistic dichotomies such as applied versus pure can be avoided. A broad participation in the hiring process can be encouraged. The advantages of hiring new people who bridge different areas can be clearly seen. In short, faculty members team to support moves which advance the department as a whole and forego the narrow and territorial perspective.

Moreover, the advantages that accrue to a department from a contract guaranteeing resources over the long run directly contribute to the mission of the university as a whole. The department is fully co-opted into providing leadership for all things mathematical that go on in the university, for the education and training of its undergraduates and graduate students, for the provision of an intellectual home for all those colleagues from different disciplines who share common mathematical challenges and intents, and for the advancement of knowledge within the discipline itself.

Chapter 18
Advice from a Department Head

John Conway[1]

A mathematics department with a graduate program has three significant areas of involvement whose combination makes it unique amongst all campus units: precalculus and calculus service courses, the program for the majors and power users of mathematics, the graduate and research programs. Promoting and fostering all three areas and getting the department to recognize the importance of all three are the keys to academic prosperity.

Some other departments, such as English, also have significant low-level service courses. English and mathematics, however, are essentially the only subjects required by every unit across the campus. Of course every unit has a major program and usually a graduate and a research program as well. No unit on campus other than mathematics, however, offers upper-division and graduate-level courses to students majoring in other disciplines.

Indeed, many mathematics departments have advanced courses populated almost exclusively by engineers and scientists. In contrast, it is a rare year that a graduate student in history takes a senior-level course in Shakespeare.

Service Courses and Calculus

The truth is that if any of these three areas of activity within the mathematics department are ignored, severe consequences are likely to follow. The service program handles so many students that any neglect here is likely to be heard all the way up the administrative food chain. But even though it handles the largest number of students, possibly double the number encountered in the other two parts combined, it cannot be allowed to become the tail that wags the dog.

Having an impact on a great number of people is certainly what this service mission does, though the other parts of the mission, in the long run, also impact large numbers. Placing supreme emphasis on servicing large numbers of students is shortsighted and inimical to the profession and the health of the department.

---

[1] John Conway is professor of mathematics at the University of Tennessee, where he also serves as head of the Department of Mathematics.

The service program, despite its size, is unlikely to obtain tenure-track positions for the department. University administrators recognize that it does not take a Ph.D. mathematician to teach such elementary courses. The department nevertheless must devote energy and resources to this part of its curriculum.

My way of maintaining quality control and improvement in the service courses is to put faculty in charge who consider this assignment as professional development. In our department we are blessed with several instructors, hired without tenure but on a somewhat permanent basis (it's a complicated story), for whom this administrative responsibility is their primary activity outside of teaching. They do a wonderful job of keeping the courses updated and coordinating the various sections of the same course to insure that the syllabus is followed and there is some degree of uniformity. They also alert me and the tenured faculty to problems and their possible solutions, as well as to developments in the approach to this material. Some of these instructors have Ph.D.'s and some do not.

Yes, I do have some tenured faculty who could do this job and would do it well. But I do not have enough of them to fill all the roles needed. For service courses it is important, however, that tenure-track faculty monitor what has been happening and what has developed in the service-level courses. In the final analysis it is the responsibility of the mathematics department to be certain that these courses are taught well. Blending all these elements together requires time and effort from many quarters. There are several pitfalls, and communication is an essential key to a smooth operation.

Having instructors and tenured faculty serve together on the undergraduate committee is one way to foster communication. E-mail chat lines about various courses is another. Having tenured faculty occasionally teach a service course and participate in the coordinating procedure conducted by an instructor are also ways to keep the tenured faculty current in the practices at the freshman level.

Now the core calculus course is an anomaly in all this. I am ambivalent whether this belongs to the service program or with the major program. Core calculus has many of the characteristics of a precalculus service course: it has many sections, many students enter calculus courses improperly prepared, it's taught in some form in the high schools, and the failure rate is higher than courses in the major program. It is, however, the starting point for the mathematics and science majors. So in practice I have chosen to treat it as part of the major program. It is therefore important that it be taught by people who understand the subsequent courses. For me this means only Ph.D. faculty, the GTA who has passed the prelims, or the occasional instructor who has a background with greater sophistication than usual.

The Program for Majors

If I were asked the primary mission of any department, whether it is mathematics or any other discipline, I would answer that it is the teaching of undergraduates. Sadly, however, the major program is an area often neglected by mathematics departments. I frequently think that this is due to the fact that faculty energy is sapped by focusing on the graduate program and/or the service program. It is just too important for the department to have a well-run program of upper-division courses for its majors and the power users, and the creative energy of the tenured faculty is paramount for this.

There are many interesting facets to the culture of mathematics. One is the concept of the undergraduate major in mathematics held by many research mathematicians. The thinking is that the program is for those destined for graduate school and perhaps future K–12 teachers so they can properly prepare future college-bound students. The word "elitist" is certainly applicable. This narrow definition of the major program is also contrary to the underlying philosophy of modern American higher education.

It seems to me that the profession is ignoring a potential source of majors. There are people who have modest mathematical ability and might major in mathematics rather than history, or sociology, or physics. We should admit the possibility that some of our majors just want a degree and will eventually have a job in which they never use their mathematics A sound degree, say a B.A. in mathematics rather than a B.S., is a definite possibility, one which we should embrace as a concept and begin to recruit undergraduates to pursue.

The best way to maintain the research mission of a mathematics department is to have a healthy major and graduate program. Deans understand arguments that we should hire additional faculty to meet increased demand in upper-division courses. Obviously people without a Ph.D. cannot teach mathematics courses to juniors and seniors. Increasing upper-division enrollments and growing the number of mathematics majors will eventually translate into additional resources.

Graduate and Research Programs

The graduate and research programs of a mathematics department are every colleague's first love. Faculty scrutinize the graduate program, they contribute to it, and they are very concerned about keeping the program up to date. This is not unique to mathematics, and the department head can rely on the faculty's attention to the organization and conduct of the graduate and research programs.

My contribution as department head in this sphere has been more to encourage faculty to dare break with the traditional approach to graduate education and contemplate innovation in a broad sense. Such change is invariably controversial. So another role of the head in this program is to maintain departmental harmony.

In many departments, including my own, graduate courses are very lightly enrolled. Frequently the number of students in these courses drops below the university's minimum. It is hard to imagine graduate enrollments increasing to the point where they would justify additional resources.

Summary

It is the combination of these three levels of activity that sets mathematics apart from the other departments. It is also this combination of duties that causes many problems and presents many opportunities. Solve these problems and you bring (relative) prosperity. Fail to solve the problems and you cause many other problems, the least of which is a lack of even relative prosperity.

Since becoming the head of a mathematics department I have learned many things, including the uniqueness of its culture. On the one hand, we teach large numbers of students. This dictates that we become very conscious of our service role and devote energy to seeing that elementary courses run smoothly, are properly staffed, and have sufficient capacity to satisfy the demand. On the other hand, the research culture breeds an attitude of isolation from the rest of the campus and diminishes the importance of teaching elementary courses. When the concept of public service is added to the mix, the possibility of chaos and conflict between the various missions is exacerbated. The ability of a department to resolve these conflicts and pursue all its various missions successfully is the key to having a department that is well received and rewarded by the university administration, that prospers, and that, most importantly, is a pleasant place in which to work.

Most mathematics departments at research universities are large enough to provide a meaningful professional life for everyone. Usually there are faculty in these departments who are interested in each of the three classes of activity: research, teaching, and public service. Mutual respect is the key. Give them all their due, live long, and prosper.

Chapter 19
Trends in the Coming Decades

Mikhael Gromov[1]

Here are a few brief remarks on possible trends in mathematics for the coming decades.

1. Classical mathematics is a quest for structural harmony. It began with the realization by ancient Greek geometers that our 3-dimensional continuum possessed a remarkable (rotational and translational) symmetry (groups O(3) and R(3)), which permeates the essential properties of the physical world. We stay intellectually blind to this symmetry no matter how often we encounter and use it in everyday life while generating or experiencing mechanical motion, e.g., walking. This is partly due to noncommutativity of O(3), which is hard to grasp. Then deeper (noncommutative) symmetries were discovered: Lorentz and Poincaré in relativity, gauge groups for elementary particles, Galois symmetry in algebraic geometry and number theory, etc. And similar mathematics appears once again on a less fundamental level, e.g., in crystals and quasicrystals; in self-similarity for fractals, dynamical systems, and statistical mechanics; in monodromies for differential equations, etc.

The search for symmetries and regularities in the structure of the world will stay at the core of pure mathematics (and physics). Occasionally (and often unexpectedly) some symmetric patterns discovered by mathematicians will have practical as well as theoretical applications. We have seen this happening many times in the past: for example, integral geometry lies at the base of x-ray tomography (CAT scan), arithmetic over prime numbers leads to the generation of perfect codes, and infinite-dimensional representations of groups suggest a design of large economically efficient networks of a high connectivity.

2. As the body of mathematics grew, it became subject to a logical and mathematical analysis. This has led to the creation of mathematical logic and

[1] Mikhael Gromov is professor of mathematics at the University of Maryland and professor at the Institut des Hautes Études Scientifiques (IHES). This essay was included as Appendix 3 of the *Report of the Senior Assessment Panel of the International Assessment of the U.S. Mathematical Sciences,* March 1998, published by the National Science Foundation.

then of theoretical computer science. The latter is now coming of age. It absorbs ideas from classical mathematics and benefits from technological progress in computer hardware which leads to a practical implementation of theoretically devised algorithms. (Fast Fourier transform and fast multiple algorithm are striking examples of the impact of pure mathematics on numerical methods used every day by engineers.) And logical computational ideas interact with other fields, such as the quantum computer project, DNA-based molecular design, pattern formation in biology, the dynamics of the brain, etc. One expects that in several decades computer science will develop ideas on even deeper mathematical levels, which will be followed by radical progress in the industrial application of computers, e.g., a (long overdue) breakthrough in artificial intelligence and robotics.

3. There is a wide class of problems, typically coming from experimental science (biology, chemistry, geophysics, medical science, etc.), where one has to deal with huge amounts of loosely structured data. Traditional mathematics, probability theory, and mathematical statistics work pretty well when the structure in question is essentially absent. (Paradoxically, the lack of structural organization and of correlation on the local level lead to a high degree of overall symmetry. Thus the Gauss law emerges in the sums of random variables.) But often we have to encounter structured data where classical probability does not apply. For example, mineralogical formations or microscopic images of living tissues harbor (unknown) correlations which have to be taken into account. (What we ordinarily "see" is not the "true image" but the result of the scattering of some wave: light, x-ray, ultrasound, seismic wave, etc.) More theoretical examples appear in percolation theory, in self-avoiding random walks (modeling long molecular chains in solvents), etc. Such problems, stretching between clean symmetry and pure chaos, await the emergence of a new brand of mathematics. To make progress, one needs radical theoretical ideas, as well as new ways of doing mathematics with computers and closer collaboration with scientists in order to match mathematical theories with available experimental data. (The wavelet analysis of signals and images, context-dependent inverse scattering techniques, geometric scale analysis, and x-ray diffraction analysis of large molecules in crystallized form indicate certain possibilities.)

Both the theoretical and industrial impacts of this development will be enormous. For example, an efficient inverse scattering algorithm would revolutionize medical diagnostics, making ultrasonic devices at least as efficient as current x-ray analysis.

4. As the power of computers approaches the theoretical limit and as we turn to more realistic (and thus more complicated) problems, we face the "curse of dimension" which stands in the way of successful implementations of numerics in science and engineering. Here one needs a much higher level of mathematical sophistication in computer architecture as well as in computer programming, along with the ideas indicated above in (2) and (3). Successes here may provide theoretical means for performing computations with growing arrays of data.

5. We must do a better job of educating and communicating ideas. The volume, depth, and structural complexity of the present body of mathematics make it imperative to find new approaches for communicating mathematical discoveries from one domain to another and drastically improving the accessibility of mathematical ideas to nonmathematicians. As matters now stand, we mathematicians often have little idea of what is going on in science and engineering, while experimental scientists and engineers are in many cases unaware of opportunities offered by progress in pure mathematics. This dangerous imbalance must be resolved by bringing more science into the education of mathematicians and by exposing future scientists and engineers to core mathematics. This will require new curricula and a great effort on the part of mathematicians to bring fundamental mathematical techniques and ideas (especially those developed in the last decades) to a broader audience. We shall need for this the creation of a new breed of mathematical professionals able to mediate between pure mathematics and applied science. The cross-fertilization of ideas is crucial for the health of science and mathematics.

6. We must strengthen the financing of mathematical research. As we use more computer power and tighten collaboration with science and industry, we need more resources to support the dynamic state of mathematics. Even so, we shall need significantly less than other branches of science, so that the ratio of profit to investment remains highest for mathematics, especially if we make a significant effort to popularize and apply our ideas. So it is important for us to make society well aware of the full potential of mathematical research and of the crucial role of mathematics in short- and long-term industrial development.

Part V

Resources

Chapter 20
How to Conduct External Reviews

The very word "review" triggers aversion: visions of a retreat into a Kafkaesque world of bureaucracy, paper trails, endless lists, and questionnaires. Where will it all end? And an even worse thought: How and where can we begin?

We all know, however, that taking stock is important. And it is particularly important in departments such as mathematics, which although sometimes large and unwieldy, are so central to the university's overall mission. Are the vision and the goals of the department consistent with those of the university? Do we do right by our undergraduates? Are they really learning what we think and hope they are learning? Are we listening to them? Are there improvements that might be made to the curriculum? Does the training and experience we give our students enable them to go on to further advanced study or get good jobs whatever their career choices might be? How is the graduate program doing? What will we do if the numbers fall? And how are our relations with other departments? Do our colleagues in other departments see us as fellow travelers on the road to discovery and a resource, or do they regard us as isolated and insular, completely immersed in a world of our own? Are we generating the kind of resources to do what we want to do? And what is it that we really want to do anyway? Are there any areas of mathematics we should be getting into? How would we like our department to look in five years' time? And most important, are we, as departmental colleagues, all on the same page?

We suspect that none of us would disagree that it is important to ask ourselves such questions from time to time. In fact, it has been the experience of the Task Force that departments which had an overall plan and a strategic view were by and large the most successful. To develop a plan, it is necessary to undergo some self-evaluation. The exercise of self-assessment focuses the mind, allows us to take note of, and take advantage of, changes and new opportunities. The exercise itself requires a little organization. It is important to begin by trying to write a self-assessment document and then asking colleagues from within or without the university to examine and scrutinize the outcome.

To begin the process, however, is sometimes difficult. Therefore, it is useful to have a stencil, a format, a plan to follow until the process at your department takes on a life of its own and the questions ask themselves. To help get started, we suggest the following self-study guidelines. They were prepared by the AMS

Committee on the Profession from self-evaluation materials used at several universities. Not all the questions will be of interest in your situation. Discard those that are irrelevant, and put in your own instead.

Many of us have found these exercises to be not only useful but absolutely necessary as a healthy check on our well being. Moreover, they can also be helpful in establishing the department's credibility with the administration. If you feel that your contributions are not being properly recognized, that you have plans worthy of investment, that you are underresourced, make the case in your self-assessment document and ask for external opinions. Such confidence has a curious effect on deans. It makes them both pleased and nervous at the same time.

THE "WHY AND HOW" OF EXTERNAL REVIEWS OF U.S. MATHEMATICS DEPARTMENTS[1]

This document is to be viewed as generic, answering the questions:

- Why should my department undergo an external review?
- How does my department and university prepare for an external review?
- How does my department conduct an external review?

Why Undertake an External Review

Some mathematics departments are required to have routine external evaluations, some departments have sporadic evaluations, while there remain many departments that have never undergone an external review. An external review requires a large effort by the department, school/college, and central administration. Thus there should be a large return for these efforts. Here is a list of potential returns from undertaking an external review.

- The process of the review will clarify the strengths and weaknesses in your curricular, research, and support programs.
- The process of the review will clarify the strengths and weaknesses of your relationships with other departments, schools and colleges, and the central administration.
- The review will establish evaluation and subsequent planning that focuses on the identification and resolution of issues that are likely to improve your mathematics department.
- The review can advertise the successes of the department to an external group of distinguished and influential mathematicians.

---

[1] These guidelines for external reviews were prepared by the AMS Committee on the Profession from documents used by several universities, including Brown University. The introductory remarks and editorial work was largely done by Ron Stern, whose contributions we gratefully acknowledge.

Preparing for and Conducting an External Review

The external review process itself requires work and resources from all levels of the university. The central administration should provide adequate funding for the external review. This includes all travel and local expenses as well as an honorarium for each external review committee member. A typical review team may consist of three or four distinguished mathematicians, and the on-site review could take only one or as many as two and a half days.[2] There are several issues to keep in mind when selecting members for the review team. The department should keep in mind that both positive *and* negative comments from an external review team are beneficial for the growth and development of the department. Thus the potential members of the review team should be distinguished mathematicians who are able to critically assess and evaluate a department and who also have the ability to express their findings.

Several months prior to the on-site visit of the external review team, this team needs to be invited and secured for service. One member should be invited as the chair of the external review team and should be assigned the responsibility of providing a written report (reflecting the external review team's on-site visit) in a timely fashion.

Your department should identify a mathematics faculty committee with staff support to assist in a self-study. This committee will be used to discuss and answer the questions posed in this self-study.

Your department should then undertake a self-study with some guidance from the dean. The goal is to prepare a written profile of the department that includes an overview of existing curricular, research and support programs, a written mission statement, and a written statement of planned future developments.

In consultation with the dean, the next step is to develop a schedule of meetings for the external review team. These meetings should include all constituencies of the department (faculty, staff, and students) and those served by the department, a walk-through of your physical facilities, and meetings with the department chair, dean, and vice-chancellors/vice-presidents for undergraduate, graduate, and research affairs, and the chancellor/president.

The external review team should be provided with a packet consisting of the departmental self-study, the tentative schedule for the on-site review, and a list of questions that they are expected to answer. This packet should be received by the external review team at least two weeks prior to the on-site visit. The chair of the external review team should be allowed to "fine-tune" the schedule and to add other people to the schedule.

At the end of the on-site visit, the external review team should present a verbal report of its findings. Thus time should be allotted in the on-site schedule to prepare for this meeting. One model is to have two presentations. The first should be a general presentation of the team's findings, including a critique and self-

---

[2] Taking the larger numbers, a department could expect to spend $2,400 in local expenses, $2,000 in airfares, and $5,000 in honoraria ($1,000 for each member and $2,000 for the chair of the team that will do all the writing of the report).

study. Those involved in the first presentation should include the external review team, dean or vice-president, the departmental chair, and the chair of the self-study team. In order to maintain confidentiality, the second exit interview should be limited to the dean or vice-president or president and the external review team.

It is then the (paid) responsibility of the chair of the external review team to provide a written report to the dean within one or two weeks.

What to Do with the External Review Team Report

The final step can consist of one of two possible actions. The first action is for the department and administration to develop and act on a plan to implement the recommendations. The other action is for the report and/or plan to sit on the shelf to collect dust until the next external review is undertaken. The bottom line is that this exercise will be important and influential if you, the dean, or the department chair make it so.

Self-Study Outline for Mathematics Departments Undergoing External Review

This document is intended as a generic framework for your department's self-study report, which will be forwarded to the senior academic administration and the external review team. Not all questions may be relevant to your department, nor should they be answered individually. These questions should be used to guide and facilitate a thoughtful and complete written discussion of your department's current situation and future plans.

You should aim for a finished document of no more than twenty typewritten pages, which should then be supplemented by appended data and other departmental documents. This self-study will be most useful if your text is interpretive and evaluative, and if it refers to the supporting documents rather than attempting to duplicate them.

A. Overview, Goals, and Recent History

- What are your department's major goals? (If you have a mission statement, please append it.) How have your goals and/or mission statement changed over recent years? How are they expected to change in the future? Include the role of graduate and undergraduate instruction, research, relationships to other academic units at your university, and community outreach.
- How is your department organized? Describe your faculty and staff administrative structures (attach an organizational chart if appropriate).
- Describe your program's history since the last external review or within the past five to seven years. In what ways has your program improved or deteriorated within this time period? How has your department addressed any issues raised by the previous review? (Attach a copy of the report of the most recent external reviews your department has undergone if such a report exists.)

- Identify three to five mathematics departments at other universities that provide targets of aspiration for your department. How does your department compare with others nationally? What evidence suggests this conclusion?

B. Faculty

- Describe briefly the profile of the faculty in terms of the areas of teaching and research expertise and their demographic characteristics.
- Describe the profile of any professional nonfaculty staff members who make significant contributions to the academic programs of your department.
- Summarize your faculty's overall strengths and weaknesses. What information has been used in identifying these strengths and weaknesses, and what other conclusions have been drawn from this information? What is the balance of scholarly depth and breadth in the faculty, and what is the balance of traditional views as contrasted with work taking place at the field's frontiers? Have there been any significant losses or additions of fields or subfields since the last external review or in the last five to seven years?
- Describe your faculty's overall strengths and weaknesses as a teaching faculty at both the graduate and undergraduate levels. How do you assess teaching performance and in what activities does your faculty participate that improve the quality of teaching in your department?
- Describe and evaluate the faculty's participation, leadership, and influence in the academic profession through such avenues as professional associations, review panels, and advisory groups.
- Describe your department's potential for responding to changing directions and new external opportunities. What indicators show the level of morale, commitment, and continuing self-improvement of your department?
- What efforts have been made to make your department more diverse with regard to gender and race/ethnicity?
- How are junior faculty mentored? How are tenure-track faculty evaluated and kept informed of their progress towards tenure?
- What is your faculty's collective view of the program's future, its desired directions, and its means for reaching its objectives? How do planning and incentives direct the program to these ends?

C. Scholarly Productivity/Creative Performance

- Evaluate the level of scholarly activity in your department, addressing the quality and quantity of your department's publications, presentations at academic and/or professional forums, and performances as appropriate.
- Evaluate the level of internal and external support for research, performance, or creative activity in your department. Is your department competing effectively for external support? Describe any deficiencies in facilities and resources which negatively affect your department's attempts to reach its research objectives.
- Describe any significant research interactions with other units at your university and with external entities (public or private). What have been the benefits of these interactions and the drawbacks, if any? How do they contribute to your department's research goals?
- Briefly describe how the research, performance, or creative activity in your department compares nationally and internationally.

D. Undergraduate Program

- Describe and evaluate the organization of and rationale behind your department's undergraduate curriculum and course offering.
- How is the undergraduate concentration organized, and why is it organized that way? What evidence is there of sufficient breadth and depth of course offerings, as well as balance among the various specialties to meet student needs and interests? Does an external accrediting body prescribe any portion of the concentration? If so, describe how the program measures up to accreditation standards, and append a copy of the most recent accreditation report.
- What introductory courses are aimed at a liberal arts education, and are the number, range, and level of these appropriate? By what standards do you evaluate the appropriateness and effectiveness of these courses?
- What courses does your department offer (if any) that primarily serve the needs of students who are concentrating on other fields or who are meeting preprofessional school requirements? Evaluate the effectiveness of these courses.
- If there are any enrollment limits on any of your courses, describe the rationale for imposing such limits, and evaluate the costs and benefits of having such limits.
- Describe the nature of your department's undergraduate curricular planning efforts. What specific efforts are made to incorporate new knowledge and areas into the curriculum? Is this generally left to individual

faculty to decide, or is the content of the curriculum reviewed comprehensively? How are proposed new course offerings evaluated? In general, what plans are under way to change or strengthen your undergraduate offerings and programs?

- Does your department substantially support or participate in multiple concentrations? What other departments actively participate? Are there sufficient teaching and advising resources to support these concentrations? Are there redundancies in these concentrations?
- What efforts are made to involve students actively in their learning through internships, undergraduate teaching assistantships, research projects, seminars, independent study? What are the criteria for honors in the concentration? Are eligible students gaining access and being attracted to your honors program?
- Describe and evaluate the organization of and rationale behind your department's allocation of teaching personnel. What percentage of your courses are covered by tenure-track or tenured faculty?
- What is the faculty teaching load in your department? How are teaching assignments determined in a way that is equitable to all faculty at the same time that quality of instruction is maintained?
- What proportion of courses in various categories are taught by full-time faculty, part-time or visiting faculty, and graduate students? If these categories of faculty are not in the right proportions, describe how and why the mix should change.
- What is the role of graduate teaching assistants in your department's instructional program? How are they selected and trained for their roles? How are they supervised and evaluated? What changes, if any, in the number of teaching assistants or in the nature of the work they perform seem warranted?
- What is the role of undergraduate teaching assistants in your department's instructional program? How are they selected and trained for their roles? How are they supervised and evaluated? What changes, if any, in the number of undergraduate teaching assistants or in the nature of the work they perform seem warranted?
- How is the quality of instruction assessed and improved in your department on an ongoing basis?
- Describe the students in the undergraduate concentration program.
- Are you attracting the number and quality of students to meet your department's needs and expectations? If not, how can changes be brought about? Please make your needs and expectations explicit.
- Explain any recent significant changes in undergraduate courses.

- Where do your undergraduate majors go, and what do they do after graduation? What indicators do you use to monitor the success of your graduates? How does the quality of the graduates compare with student quality in your field nationally? How do alumni of your program view their educational experience? Describe any honors or awards received by undergraduate concentrators.
- Describe and evaluate the process and structure of your undergraduate advising.
- Describe the nature of and evaluate any outreach activities in your academic department that impact on undergraduate education.

E. Graduate Program

Overview

- Describe, in general terms, the graduate program(s) offered by your department. What changes have occurred in recent years, and what changes are contemplated for the future?
- What evidence (e.g., reputation, recruiting and retention, outcomes) is available concerning the quality of your department's graduate program(s)? How is the information used to strengthen the graduate program(s)?

Curriculum and Courses

- What evidence is there of sufficient course and research opportunities and balance among the various specialties? How are the courses in your graduate program coordinated?
- Do students have adequate resources to carry out their studies (e.g., office and lab space, supplies, travel, library collections, and financial support)? What additional resources would be required to improve the quality of your graduate program substantially?
- Does your department offer graduate courses taken by significant numbers of students from other programs? Does your department depend upon courses offered by other units? Describe the planning process used for these courses, how the offerings are coordinated with the other units (including coordination problems encountered), and how well the courses meet the needs of all programs involved.

Graduate Students

- What mechanisms are used to recruit students? Is the program competing well for top students? What help is needed in recruiting? How does the quality of students in your graduate program compare with student quality in other similar programs? How does the quality and quantity of current students compare to the students in your program five years ago? Ten years ago?

- What is the current gender and race/ethnicity composition of graduate students in your program? How do these figures compare with similar figures for undergraduates? For graduate programs at other schools? What efforts are under way to attract and retain well-qualified students from nonmajority groups?

Professional Training, Advising, Placement

- Describe and evaluate the preliminary/qualifying examination requirement in your program(s).
- In what ways, besides individual thesis or dissertation research, do graduate students receive professional experience (e.g., research assistantships, internships, outreach efforts, etc.)?
- How do graduate students acquire professional skills other than those directly associated with research and teaching (e.g., learning how to write grants, give colloquia, etc.)?
- What is the nature and quality of the advising for graduate students, and how is such advising assessed?
- How well do your master's and Ph.D. students fare on the academic job market? On the nonacademic job market? How is placement information used to evaluate and modify the nature of the graduate program?

F. Administration and Support Services

- Describe and appraise any related support activities that impact your teaching, research, and/or service programs (e.g., outreach efforts).
- Describe and appraise the physical facilities associated with your department.
- Describe and appraise the current levels and types of staff support (both technical and office).
- Rank order you department's specific and most pressing support needs (for example, library, computer equipment/support, office personnel, technical assistance, etc.).

G. Summary Assessment and Future Directions

In no more than two pages, highlight the most salient points of this self-study. Place emphasis on plans, new directions, and remediation of existing problems and on ways your department is working to help itself.

Materials to Be Appended

- Department Profile (e.g. number of faculty, budget dollars, grant dollars, etc.) and area comparative data (to be provided by the dean).
- Graduate program data (to be provided by the graduate school).
- Mission statement.
- Copy of the report of the most recent external review committee.
- Faculty and staff administrative organization charts.
- CV's of all regular faculty and staff members that have regular teaching responsibilities at the graduate and/or undergraduate level.
- A copy of all departmental informational publications, including graduate and undergraduate program descriptions, graduate manual, the "department brochure" (if there is one), etc. Include any newsletters to graduate or friends of your department.

Chapter 21
Where to Find Data (and How to Use It)

How do you convince a distrustful dean that your declining number of majors is part of a national trend? How can you plan for changing applications to the graduate program? How do you prepare your doctoral students for the job market they face in the next few years? The answers to all these questions begin with data.

Accessing and understanding data is a key part of making a convincing argument. But it is also a key to planning for a department's future and understanding the environment in which it exists. If you want to expand a program that has successfully increased the number of majors, show the administration that you have reversed a national trend. Making arguments with carefully prepared data helps to convince the listener for two reasons: the facts themselves, and the fact that you gathered them. But data is equally useful for planning and understanding. Department chairs (and others in the department's leadership) need to know what major national trends are affecting their discipline, and they need to compare their own situation to those trends. Much understanding comes from that process.

Four key sources of data are listed below. The sources range from data specific to U.S. departments of mathematics to data on the national science and engineering enterprise. These sources often point you to other sources of related data.

Resources for Data on Mathematics in Academia

1. The jointly sponsored *Annual Survey of the Mathematical Sciences*, newly renamed with the 1998 survey cycle, collects data from academic departments in the mathematical sciences and from each year's doctoral recipients. Regular data collection efforts were begun by the AMS in 1957. MAA became a joint sponsor of the modern survey effort in 1989, IMS in 1993, and ASA in 1998. The survey currently gathers information annually on faculty salaries and counts of faculty by rank and sex, total and first-year counts of graduate students, undergraduate and graduate enrollments, and junior/senior majors. It also gathers information on doctoral recipients and their initial employment experiences, first from the doctoral-granting departments, then from the individual recipients in a follow-up survey.

Reports on the Annual Survey of the Mathematical Sciences are published periodically in the *Notices* of the AMS. A complete list of reports for the past five years may be found at http://www.ams.org/membership/survey.html, with links to those available electronically. Reprints are also available from the AMS by phone (401-455-4113) or e-mail (survey@ams.org).

2. The *Conference Board on the Mathematical Sciences Survey on Undergraduate Instruction in Two- and Four-year Institutions* has been conducted every five years since 1965. This survey gathers detailed enrollment data by individual course for the whole range of undergraduate mathematics and statistics courses. It also gathers counts of faculty by age, rank, sex, and, in recent years, race/ethnicity. The strength of this survey is its long-term trend data on undergraduate instruction. Each survey also includes a section of questions on topics of then-current interest.

Copies of the fall 1995 CBMS survey report, the most recent in the series, were mailed in June 1997 to all mathematics and statistics departments. Additional copies may be purchased from the MAA by calling 800-331-1622. An overview of the entire survey is available on e-MATH at

http://www.ams.org/membership/survey.html.

Data on How Mathematics Fits into Science and Engineering

3. The Science Resources Studies (SRS) Division of the National Science Foundation is the unit that manages the survey efforts supported by NSF. It is by far the richest source of data comparing the mathematical sciences with other science and engineering disciplines. Access to the reports on these surveys has been made much easier by the World Wide Web. The starting point for information available through SRS is

http://www.nsf.gov/sbe/srs/stats.htm.

Two of the valuable long-standing surveys managed by SRS are the annual Survey of Earned Doctorates and the longitudinal Survey of Doctoral Recipients. The first of these overlaps with the AMS Annual Survey, but it becomes available considerably later and does not provide as detailed a look at new doctoral recipients in mathematics as does the Annual Survey. Its advantage is that it provides comparable data for all the science and engineering disciplines, e.g., a measure of time-to-degree. The Survey of Doctoral Recipients is a longitudinal sample survey of the complete population of U.S. doctoral recipients over the past fifty years. It is designed to provide demographic and career history information about individuals with U.S.-granted doctoral degrees. A third survey of particular interest is the Graduate Students and Postdoctorates in Science and Engineering (GSS) survey, which obtains data on the number and characteristics of graduate science and engineering (S&E) students enrolled in U.S. institutions. The results of the survey are used to assess trends in financial support patterns and shifts in graduate enrollment and postdoctorates.

4. Another source of general data on postsecondary education is the National Center for Education Statistics (NCES), a unit of the Office of Educational Research and Improvement of the U.S. Department of Education. NCES maintains a number of ongoing surveys of postsecondary education, including detailed data

on undergraduate and graduate enrollments, degrees awarded, and staffing in postsecondary institutions. NCES's Web site,

http://nces.ed.gov/surveys/datasurv.html,

provides easy access to its many reports of these surveys (via PDF files), as well as access to public-use data sets. Some of this data is available for certain disciplines, including mathematics. A report on the view of mathematics enrollments provided by this data will be published by the AMS in the future.

The following examples illustrate the kinds of issues a department chair might encounter for which data is available from the sources above.

• What has been happening to undergraduate enrollments?

The enrollment in undergraduate mathematics courses within mathematics departments at four-year institutions declined 9% between fall 1990 and fall 1995, from 1,619,000 to 1,469,000. This decline in four-year institutions contrasts with a 12% increase in mathematics enrollments at two-year institutions, from 1,241,000 to 1,384,000. The fall 1995 enrollments at two-year institutions accounted for 49% of the total enrollment in undergraduate mathematics taught within mathematics departments. (From 1995 CBMS highlights by Donald Rung, *Notices of the AMS,* vol. 44, no. 8(Sep 1997), 923–931.)

• What has been happening to enrollments in graduate courses?

Total enrollment in graduate courses in Ph.D.-granting mathematics departments declined 19% between fall 1992 and fall 1997, based on enrollments data collected in the AMS-IMS-MAA Annual Survey.

Fall Graduate Course Enrollments, 1992 to 1997

	Fall 1992		Fall 1997		
Departmental Groupings	**Count**	**% of Total Enrollment**	**Count**	**% of Total Enrollment**	**% Change 1992 to 1997**
Group I Public (19 of 25 responding)	6,892	33.0%	4,964	29.2%	-28.0%
Group I Private (14 of 23 responding)	2,101	10.1%	1,943	11.4%	-7.5%
Group II Public (33 of 44 responding)	6,361	30.5%	5,447	32.0%	-14.4%
Group II Private (6 of 12 responding)	556	2.7%	397	2.3%	-28.6%
Group III Public (31 of 51 responding)	4,339	20.8%	3,763	22.1%	-13.3%
Group III Private (9 of 21 responding)	610	2.9%	486	2.9%	-20.3%
Total enrollment (112 of 176 responding)	**20,859**		**17,000**		**-18.5%**

Table 1

The decline within the Group I Public departments was an even more dramatic 28%, while the decline within the Group I Private departments was a much less dramatic 8%. The above table is based on an unpublished retrospective analysis of data provided by the 112 Ph.D.-granting mathematics departments that responded to both the 1992 and 1997 Departmental Profile survey, one of four surveys that comprise the Annual Survey. These 112 departments account for 70% of the Ph.D.'s produced over the last ten years by the 176 departments in Groups I–III.

• What has been happening to numbers of graduate students?

The number of full-time graduate students in Ph.D.-granting mathematics departments declined 19% between fall 1991 and fall 1997.

Counts of Full-Time Graduate Students, 1991–1997

	Female			Male			Total		
	1991	1997	% Change	1991	1997	% Change	1991	1997	% Change
Group I Public (19)	624	507	-19%	1,981	1,514	-24%	2,605	2,021	-22%
Group I Public (Top 9)	370	296	-20%	1,296	920	-29%	1,666	1,216	-27%
Group I Private (17)	175	145	-17%	687	549	-20%	862	694	-19%
Group I Private (Top 8)	71	71	0%	373	308	-17%	444	379	-15%
Group II (39)	685	608	-11%	1,587	1,288	-19%	2,272	1,896	-17%
Group III (37)	432	399	-8%	854	711	-17%	1,286	1,110	-14%
All Departments (112)	1,916	1,659	-13%	5,109	4,062	-20%	7,025	5,721	-19%

Table 2

The decline within nine of the top twelve Group I public departments for which data was available was 27%, while it was 22% for the group as a whole. The declines were 19%, 17%, and 14% within Group I Private, Group II, and Group III respectively. Not surprisingly, the decline in first-year (full-time) graduate students was even more precipitous. Between fall 1991 and fall 1997 the decline within Ph.D.-granting mathematics departments was 26% while within Group I Public, Group I Private, Group II, and Group III departments they were 34%, 41%, 12%, and 28% respectively. These figures, taken from Tables 2 and 3 are based on a retrospective analysis of data provided by the 112 Ph.D.-granting mathematics departments that responded to both the 1991 and

1997 Departmental Profile surveys. (These responding departments differ slightly from those in Table 1.)

Counts of Full-Time First-Year Graduate Students, 1991–1997

	Female			Male			Total		
	1991	1997	% Change	1991	1997	% Change	1991	1997	% Change
Group I Public (19)	191	121	-37%	450	305	-32%	641	426	-34%
Group I Public (Top 9)	101	63	-38%	280	182	-35%	381	245	-36%
Group I Private (17)	50	38	-24%	172	92	-47%	222	130	-41%
Group I Private (Top 8)	20	18	-10%	83	48	-42%	103	66	-36%
Group II (39)	229	211	-8%	425	365	-14%	654	576	-12%
Group III (37)	173	135	-22%	308	209	-32%	481	344	-28%
All Departments (112)	643	505	-21%	1,355	971	-28%	1,998	1,476	-26%

Table 3

• What has been happening to tenure-track positions?

The total number of tenured faculty in mathematics departments remained almost constant between 1990 and 1996, around 13,400. Over this same time interval, the number of individuals in tenure-eligible positions, i.e., tenure-track but not yet tenured, declined by almost 30%, from approximately 4,700 to 3,300. In Group I, II, and III combined (the Ph.D.-granting mathematics departments) there was a 27% decline, from 1120 to 820. Figure 1 shows the recent trends in tenure-eligible positions and non-tenure-eligible positions for Groups I–III combined and Groups M and B combined. (See "Changes in Mathematics Faculty Composition", Fall 1990 to Fall 1996, *Notices of the AMS*, **44**, no. 10 (Nov 1997), 1321–1323.)

• What is the situation relative to the use of part-time faculty in mathematics departments?

In the Ph.D.-granting departments, the number of individuals holding part-time appointments increased slightly between 1990 and 1996, from 975 in fall 1990 to 1,090 in fall 1996. In master's and bachelor's departments, the numbers of individuals in part-time appointments declined slightly, from 5,200 in fall

1990 to 4,930 in fall 1996. (See "Changes in Mathematics Faculty Composition", Fall 1990 to Fall 1996, *Notices of the AMS,* **44**, no. 10 (Nov 1997), 1321-1323.)

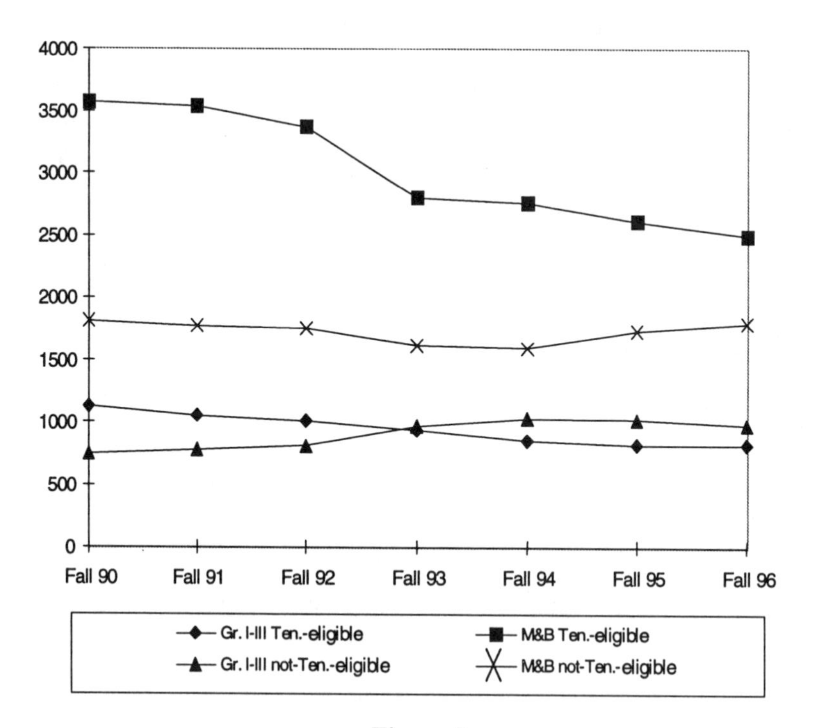

Figure 1

• What were the starting salaries for postdoctoral positions in fall 1997?

The median salary for a 9–10-month appointment for a postdoctoral position in academia in fall 1997 was $38,500. A report on starting salaries for new doctoral recipients is a regular feature of the Annual Survey reports. Starting salaries for fall 1997 appear in the "Second Report of the 1997 AMS-IMS-MAA Annual Survey", *Notices of the AMS,* **45**, no. 9 (Oct 1998), 1163–1165.

Finally, a cautionary note is in order. Gathering massive amounts of data and then using it to support whatever arguments one proposes can be counterproductive. (A famous quote of Andrew Lang goes: "He uses statistics as a drunken man uses lamp posts—for support rather than illumination.") There are some vital aspects of a department's life that are not easily measured by numbers. Nonetheless, data can be an important part of understanding.

Chapter 22
A Digest of Some Reports

Index of Report Summaries

This book is only one of many attempts to address the issues of research, education, and the role of mathematics. While it is difficult to be knowledgeable about all such material, mathematicians can profit by knowing what has been said, even when they disagree with it.

This chapter contains descriptions of a sample of such reports, selected to represent the variety of material. The reports include:

- 1945 Science—The Endless Frontier (Vannevar Bush Report)
- 1984 Renewing U.S. Mathematics (David Report)
- 1991 Moving Beyond Myths (MSEB)
- 1992 Educating Mathematical Scientists (Douglas Report)
- 1994 Recognition and Rewards in the Mathematical Sciences (JPBM)
- 1994 Talking about Leaving
- 1995 SIAM Report on Mathematics in Industry
- 1996 Shaping the Future, New Expectations for Undergraduate Education
- 1998 Reinventing Undergraduate Education (Boyer Report)
- 1998 Senior Assessment Panel Report
- 1998 Unlocking Our Future (Ehlers Report)

The following chapter contains a more comprehensive bibliography, with brief annotations for much of the material.

Science, The Endless Frontier — A Report to the President on a Program for Postwar Scientific Research

- *By Vannevar Bush, Director Office of Scientific Research and Development, Washington, DC, 1945 (reprinted by the National Science Foundation in 1990)*

The Office of Scientific Research and Development was established in June 1941 to coordinate weapons development and related research during World War II. It was directed by Vannevar Bush, an electrical engineer. The OSRD oversaw the development of the atomic bomb, advances in microwave radar, and mass production of penicillin. Much of its scientific work was performed at universities under contract with the government. More than fifty universities received contracts of over a million dollars each. Such levels of government support of research were unprecedented. (The four or five largest university departments of physics, chemistry, and biology each spent thirty to forty thousand dollars annually on research before the war.)

In November of 1944, with the end of the war in sight, Roosevelt wrote a letter to Bush (at least in part at Bush's instigation), asking for his recommendations on the continuation of government involvement with science. He asked in particular about four points: first, the diffusion of scientific knowledge arising from the war effort; second, the continuation of medical research undertaken for the war; third, government aid to research by public and private organizations (primarily military laboratories and universities); and fourth, the discovery and development of scientific talent.

Bush's response, delivered to Truman in July 1945, is a very precise and personal vision, contained in less than forty pages. He appointed and consulted advisory committees for each of the four points, and their reports provide a hundred and fifty pages of appendices. The answer to the central question of continued government support for science was implicit in Roosevelt's letter: "The information, the techniques, and the research experience...should be used in the days of peace ahead for the improvement of the national health, the creation of new enterprises bringing new jobs, and the betterment of the national standard of living." Bush's report echoed this theme, with emphasis on how much remained to be done: "But without scientific progress no amount of achievement in other directions can insure our health, prosperity, and security as a nation in the modern world." The report was released to the public on July 19, three days after the Trinity test in Alamogordo.

The heart of Bush's vision was a National Research Foundation, to be controlled by a board of nine civilian scientists appointed for four-year terms by the president. The Foundation was to be organized into five divisions: Medical Research, Natural Sciences, National Defense, Scientific Personnel and Education, and Publications and Scientific Collaboration. This foundation would distribute all federal support in these fields. (The Division of National Defense was to be charged with "long-range scientific research on military matters." The military would retain direct control over "research on improvement of existing weapons.")

At the same time, Senator Harley Kilgore of West Virginia was proposing a slightly different vision: a National Science Foundation, structured more like other federal agencies and including more direct consideration both for the social sciences and for social goals. Kilgore, for example, wanted at least part of the federal support for research to be distributed on a geographic basis, and he wanted to include guarantees that small businesses could enjoy some of the benefits of technology developed with government support.

The political dispute between Bush's vision and Kilgore's lasted five years. In 1947 Congress passed a bill close to Bush's model. Truman vetoed it, saying that he could not approve an executive agency so far beyond the control of the chief executive. A compromise proposal was enacted in 1950, creating the National Science Foundation. By that time federal support for medical research was well established in the National Institutes of Health, and never moved to the NSF. Research in the natural sciences and military research were being (generously) supported by the Office of Naval Research and the Atomic Energy Commission, and both the military and the academic scientists they supported vehemently opposed any transfer to the NSF. For all of these reasons, the National Science Foundation did not become the unique center for federal support of research—the peacetime OSRD—that Bush envisioned.

Renewing U.S. Mathematics—Critical Resources for the Future

- *Report of the Ad Hoc Committee on Resources for the Mathematical Sciences, Edward E. David, Jr., Chairman, National Academy Press, Washington, DC, 1984*

In 1981 the National Research Council established the Ad Hoc Committee on Resources for the Mathematical Sciences and gave it the charge of reviewing the health and support of mathematics research in the U.S. The committee's report, widely identified as "The David Report", was published in 1984.

"The David Report" was a wakeup call to the mathematical sciences research community, professional organizations, universities, and federal agencies concerning fifteen years of deteriorating support for mathematics research. The problems were recognized earlier and, in fact, motivated the emphasis in the committee's charge on issues of support.

During the same timeframe that the Committee on Resources was working, other groups were also addressing aspects of the crisis in support for basic research. In 1982 the Committee on Science, Engineering, and Public Policy (COSEPUP) reported to the White House Office of Science and Technology Policy (OSTP) and to the Department of Defense (DoD) on research areas within mathematics that were likely to return the highest scientific dividends as a result of planned increases in federal research funding. The COSEPUP report also painted the grim picture of the status of federal support for mathematics in 1982.

"The David Report" is systematic and thorough in developing three topics:

- ♦ The strength of the mathematics research enterprise in the U.S. and the opportunities for new achievements.
- ♦ The status of federal support of the mathematical sciences and an analysis of how the crisis was allowed to happen.
- ♦ An estimate of the amounts needed for future support of research in the mathematical sciences and a plan for attaining the needed levels of support.

Appendices to the report provide data documenting the deterioration of support for mathematics and an essay by Arthur Jaffe amplifying the strength, achievements, and opportunities in mathematics perceived at the time.

The report's assessment of support in the early 1980s emphasizes how the impact of changing patterns of support over the preceding period had a more drastic impact on mathematics than on other disciplines.

"Since the late 1960s, support for mathematical sciences research in the United States has declined substantially in constant dollars, and has come to be markedly out of balance with support for related scientific and technological efforts.

"...We estimate the loss in federal mathematical funding to have been over 33% in constant dollars in the period 1968–73 alone; it was followed by nearly a decade of zero real growth, so that by FY 1982 federal support for mathematical sciences research stood at less than two thirds its FY 1968 level in constant dollars."

All sciences were affected by the changed federal policies for support of graduate students starting in the late 1960s and by the 1969 Mansfield Amendment. NSF graduate fellowships were sharply curtailed, and NDEA fellowships disappeared. The Mansfield Amendment to the FY 1970 Military Procurement Authorization limited research sponsored by the Defense Department to studies and projects that directly and apparently related to defense needs, or mission relevance. Before enactment of the amendment, the defense agencies provided substantial support for basic research in mathematics.

In 1971 and 1972 Congress increased NSF appropriations substantially to help provide for the shift from DoD to NSF of support for basic research. However, the $50 million increase for NSF did not help mathematics. At the time, as a matter of federal policy, there was greater emphasis on areas connected to industrial development such as chemistry and materials research. While the NSF budgets for support of both chemistry and physics increased at average annual rates of 20% from 1970 to 1972, the average annual increase for mathematics was a meager 4.7%.

Also in the early 1970s NSF worked to fill the void left by the shrinking or disappearance of the NSF and NDEA fellowship programs. Funds were made available through the research budgets for more research assistantships. The budgets for chemistry and physics show the positive effects of this funding.

However, mathematics did not garner the same increases. In part this resulted from inaction by the mathematics research community and its concern about a possible oversupply of new Ph.D.'s; the community did not make a strong case for the increase in NSF funding for research assistantships, and that support flowed in other directions.

By 1982 the NSF budget for support of mathematics research had actually declined in constant dollars from its 1968 level. Over the same period, the NSF budgets for chemistry and physics research had both grown on the order of 25% in constant dollars.

"The David Report" set goals for rectifying the support problems and recommended actions by the federal government, universities, and the research community. The goals included:

- support for 1,000 graduate students actively doing research for the Ph.D.,
- support of 200 new multiyear postdoctoral fellowships annually,
- support of 400 research grants for young investigators, and
- research funding for at least 2,600 established (senior) mathematical scientists.

Most significantly, the report captured the attention of the research community. The community as a whole, including professional organizations and federal agencies, worked toward its goals.

"The David Report" fifteen years later is still an important and timely document for the mathematics research community. The imbalances in support that developed between 1968 and 1982 have not been erased, even though significant progress has been made since the report. The forces that were felt in the 1970s recur, and the messages sent to the research community, professional societies, universities, and the federal agencies are worth remembering as we continue to address issues of the health and support of mathematics research.

Moving Beyond Myths—Revitalizing Undergraduate Mathematics

- *Committee on the Mathematical Sciences in the Year 2000, Board on Mathematical Sciences, Mathematical Sciences Education Board, National Research Council, National Academy Press, Washington, DC, 1991*

"Moving Beyond Myths" was written by the Committee on the Mathematical Sciences in the Year 2000, under the aegis of the Board on Mathematical Sciences, the Mathematical Sciences Education Board, and the National Research Council. It complements the booklets *Renewing U. S. Mathematics*, *Everybody Counts*, and *A Challenge of Numbers*.

"Moving Beyond Myths" is a 1991 critique of U.S. undergraduate mathematics education. It lists several myths pervading thc public perception of mathematics. These myths include:

- Success in mathematics depends more on ability than on hard work.
- Women and members of certain ethnic groups are less capable in mathematics.
- Most jobs require little mathematics.
- All useful mathematics was discovered years ago.
- To do mathematics is to calculate answers.

Furthermore, MBM says the U.S. colleges and universities perpetuated, if not created, these myths with their attitude toward undergraduate teaching.

From 1970 to 1990 mathematics enrollments increased by more than 70%, while faculty size increased by less than 30%; and instead of forcefully articulating the need to maintain low student-faculty ratios, mathematics departments acquiesced to their increased workload by teaching ever larger classes and by putting less prepared graduate assistants and part-time teachers in the classrooms. By 1990 the system was "beset on all sides by inadequacies and deficiencies":

- in the mathematical preparation of students,
- in rewards and support for teaching,
- in teaching innovations,
- in the use of computers in undergraduate mathematics,

and many others, such as a shortage in the number of mathematics students (graduate, undergraduate, women, and minority), and in the number of qualified school mathematics teachers.

"Moving Beyond Myths" also criticizes the large reliance on teaching via lectures, which "place students in a passive role," the irreverence of mathematics courses to the majority of the students' future needs, and the general effect upon students of a professional value system that rewards research more than teaching.

"Moving Beyond Myths" then challenges the mathematical community to "restructure fundamentally the culture content, and context of undergraduate mathematics education," and lists four goals:

1. Effective undergraduate mathematics instruction for all students
2. Full utilization of the mathematical potential of women, minorities, and the disabled
3. Active engagement of college and university mathematicians with school mathematics, especially in the preparation of teachers

4. A culture for mathematicians that respects and rewards teaching, research, and scholarship

"Moving Beyond Myths" offers an Action Plan, with recommendations for faculty, mathematics departments, colleges and universities, professional societies, and the government. Recommendations for faculty include:

- Learn about learning; explore alternatives to "lecture and listen".
- Involve students actively in their learning.
- Teach future teachers in the ways they will be expected to teach.
- Teach the students you have, not the ones you wish you had.

Recommendations for departments include:

- Assign the best teachers to introductory courses.
- Use knowledge gleaned from minority projects.
- Build a team of faculty to carry out experiments.
- Have a departmental seminar on issues of teaching and learning.
- Employ varied instructional approaches.

Written in 1991, "Moving Beyond Myths" is somewhat out of date (for example, with its call for computerization), but it remains a valuable guide for chairs, departments, and faculty. Articulately written, it contains several valuable examples.

Educating Mathematical Scientists: Doctoral Study and the Postdoctoral Experience in the United States

- *Committee on Doctoral and Postdoctoral Study in the United States, Board on Mathematical Sciences, Commission on Physical Sciences, Mathematics, and Applications, National Research Council, National Academy Press, Washington, DC, 1992*

This 1992 report continues to be a highly valuable resource for any mathematics department looking for suggestions for enhancing its doctoral program. The report was prepared by the Committee on Doctoral and Postdoctoral Study in the United States of the NRC Board on the Mathematical Sciences. The chair of the committee and chief author of the report was Ron Douglas, now provost at Texas A&M. The committee based its findings on site visits to a diverse set of programs in ten universities, small and large, public and private, geographically diverse; four departments were ranked in the "Top 20".

The report anticipates several subsequent NRC reports in arguing for broader doctoral and postdoctoral training to prepare Ph.D.'s for a variety of nonacademic jobs. It contains helpful suggestions about general issues, such as recruiting and retaining doctoral students, and specific issues, such as placing foreign students with weak language skills but advanced training in mathematics.

The report seeks to characterize the best practices of doctoral and postdoctoral education in the United States, a world leader in mathematical sciences research and in doctoral and postdoctoral education. The committee was looking for programs that accomplish the following two objectives.

- All students, especially the majority who will spend their careers in collegiate teaching, government laboratories, business, and industry, need to be well prepared by their doctoral and postdoctoral experience for their careers.
- Larger percentages of domestic students and, in particular, women and underrepresented minorities need to be attracted to the study of and careers in the mathematical sciences.

The committee's findings were meant to respond to growing concerns that many doctoral students are not prepared to meet undergraduate teaching needs, establish productive research careers, or apply what they have learned in business and industry. The inadequate preparation, high attrition, declining interest of domestic students, particularly women, and the near-absent interest of students from underrepresented minorities in doctoral study were problems in the early 1990s, and they are likely to remain problems into the next century.

The report suggests that even with limited resources, a successful doctoral program can flourish if, among other things, the mathematics department focuses its energies rather than trying to implement a "standard" or traditional program that covers too many areas of the mathematical sciences. It also notes that departments with the best faculty do not necessarily have the most successful doctoral and postdoctoral programs.

In its site visits, the committee conducted in-depth interviews with students, faculty, and administrators. It looked for features that were present in successful programs as well as for elements that were detrimental to quality education. The committee noted that successful programs possessed, in addition to the sine qua non of a quality faculty, the following three characteristics:

- A focused and realistic mission, with clearly defined goals and adequate "human and financial resources" to meet those goals;
- A positive learning environment, where students receive assistance, nurturing, feedback, and encouragement in a cordial atmosphere;
- Provision for relevant professional development, i.e., a program tailored to the career objectives of the students, whether undergraduate teaching, academic research, or work in government laboratories, industry, or business.

The committee identified two kinds of models for programs:

1. The standard model, which supports research in a broad range of areas, with depth in each one, and has as its goal the preparation of talented, well-motivated doctoral students and postdoctoral associates for careers as mathematical scientists at research universities.

2. The specialized models, such as the subdisciplinary model, the interdisciplinary model, the problem-based model, and the college-teachers model, which were seen to alleviate two large, human resource problems: difficulty in recruitment and replacement; and the desirability of clustering of faculty, postdoctoral associates, and students—a practice that helps create a positive learning environment and promotes relevant professional development.

Both standard and specialized programs can be successful. However, programs that do not have the human or financial resources to run a successful standard program should consider whether a specialized model might better fit their needs.

The Standard Model. The report describes the shortcomings of the American standard doctoral and postdoctoral programs. It suggests that most standard programs do well in preparing their best students and postdoctoral associates for the academic research job market, but very few prepare any of their students well for jobs in teaching, government, business, or industry. It also suggests that some of these programs struggle because they cannot attract the graduate students necessary to function as a standard-model program. The committee acknowledged the continuing need for well-established standard programs at a small number of centers and encouraged efforts to broaden the experience of students in those programs and to provide a more supportive learning environment.

The Subdisciplinary Model. For subdisciplinary models in both pure and applied areas, the department concentrates much of its faculty and resources in a few subdisciplines of the mathematical sciences. Recruiting strong, well-prepared students for subdisciplinary programs requires considerable effort to ensure a proper fit. The main advantage of the subdisciplinary model is that clustering of students and faculty working on related topics enables them to assist each other in their common goals. Some of the doctoral programs with the best reputations for research are subdisciplinary programs.

The Interdisciplinary Model. The interdisciplinary program is usually only one among several programs in a department of mathematics, statistics, or operations research. It utilizes department faculty with interdisciplinary interests and mathematically oriented faculty in cognate disciplines. The curriculum, which often involves course work in one or more other departments in science or engineering, trades depth in the mathematical sciences for greater breadth overall. Students can choose thesis advisers from the mathematical sciences department or one of the other departments. Faculty in both departments often adopt a cooperative approach to directing Ph.D. research. Graduates of interdisciplinary programs sometimes move into other disciplines or take positions in industry. These programs succeed in bringing mathematically well-trained students into fields in which they can effectively use their talents and at the same time promote the transfer of mathematical knowledge to these fields.

The Problem-Based Model. In a problem-based model, a specific application or set of applications is used as a unifying theme for courses and research. The program is concerned with the strictly mathematical aspects of an applied program, and mathematical modeling is a common focus. An attraction of such programs is that the students are immersed in research-related activities from the beginning. Student internships in regional industries are often an integral part of the program. Industrial researchers often visit the students and faculty of the program. Post-Ph.D. employment opportunities in industry are common, but graduates also obtain positions in academia.

The College-Teachers Model. Designed specifically to prepare teachers for two- and four-year college employment, this model is to be distinguished from a program that confers doctor of arts and doctor of education degrees. Breadth of course work, an emphasis on professional development in pedagogy, and a research apprenticeship are parts of the program. Most new Ph.D.'s from standard programs currently take jobs in college teaching but are often ill prepared for teaching. New Ph.D.'s from a college-teachers program are attractive candidates for employment because they are prepared to be teachers.

The committee noted the following common features of specialized models:

- Students in specialized programs find it easier to obtain appropriate jobs than do those in standard programs.
- A smaller department is more likely to be successful if it adopts one of the specialized models.
- Recruiting of domestic students, as well as women and minorities, is more effective for specialized programs than for standard programs.

The report has the following general recommendations.

- New Ph.D.'s with a broad academic background and communication skills appropriate for their future careers are better able to find jobs.
- Active recruiting increases the pool of quality students; it does not just reapportion the pool. It also increases the number of women and underrepresented minorities. Students with strong mathematical backgrounds have a choice of studying mathematical sciences, physical sciences, engineering, law, medicine, and other areas. More of them can be attracted to the mathematical sciences.
- Clustering faculty, postdoctoral associates, and doctoral students together in research areas is a major factor in creating a positive learning environment.
- A positive learning environment is important to all doctoral students, but is crucial for women and underrepresented minorities.
- All departments, including those characterized as elite and selective, need to provide a supportive learning environment.

- Doctoral students and postdoctoral fellows should receive broad academic preparation appropriate for their future careers in research universities, teaching universities, government laboratories, business, and industry.
- Doctoral students and postdoctoral fellows should learn teaching skills and other communication skills appropriate for their future careers.
- The number of postdoctoral fellowships in the mathematical sciences should be greatly increased so that such positions can be viewed as the logical next step after completion of the doctorate for the good student, not as a highly competitive prize for a select few. More postdoctoral fellowships should have applied, interdisciplinary, or pedagogical components. (Note: This report played a major role in changing the name used for initial visiting faculty positions for new Ph.D.'s in research mathematics departments from "instructor" to "postdoc".)

This summary draws heavily from text in the Executive Summary of this BMS report and the article about it by Ed Block appearing in the May 1992 *SIAM News*.

Recognition and Rewards in the Mathematical Sciences

- *Report of the Joint Policy Board for Mathematics, Committee on Professional Recognition and Rewards, American Mathematical Society, Providence, RI, 1994*

A committee of thirteen mathematical scientists, chaired by Calvin Moore, was charged with investigating the current reward systems at a wide variety of mathematical sciences departments, initiating dialogues about the issues raised, and making recommendations for improvements. The committee made site visits to twenty-six institutions of all types, convened focus group discussions at several professional meetings, and conducted a broad survey of opinion from faculty members and department chairs about key issues. This survey sampled opinion not only on what is current practice but also on what those surveyed felt "should be" the practice.

There are several categories of rewards and recognition: some are direct, such as salary, promotion, and tenure; some are more indirect, such as sabbaticals, awards for outstanding teaching, grants, course release for special projects, etc; and some are less tangible "quality-of-life" issues, such as collegiality within a department.

The committee found widespread dissatisfaction about the current reward systems, including a concern that research is overemphasized, a lack of flexibility to accommodate changing contributions throughout faculty careers, and much discomfort about the evaluation of teaching and service. The surveys also revealed a significant disparity between how chairs and faculty viewed certain issues, although this disparity was not observed during the site visits. For example, only 28% of faculty at the top-ranked 39 doctoral departments felt that salaries reflect differences between excellent and average teaching, while 55% of chairs felt this to be true.

The committee arrived at ten findings and three guiding principles, described below. Each finding is supported by data from the surveys and information from the site visits and focus groups. Although the committee did not reach any sweeping recommendations, in part because of the diversity of the institutions involved, it did suggest that "The recognition and rewards system in mathematical sciences departments must encompass the full array of faculty activity required to fulfill departmental and institutional missions." The report concludes with an appendix on "Defining Mathematical Scholarship".

Findings

1. There is a substantial gap between what faculty members think the rewards structure should be and what it actually is, as well as a desire for a broader and more flexible rewards structure.

2. During the last five to ten years there has been an evolution in mathematical sciences departments, with an increased emphasis on research and scholarship in the departments which traditionally emphasized their teaching roles, while at the same time there has been an increased emphasis on the teaching roles in departments which traditionally emphasized their research roles.

3. Survey results from questions about the importance of three different types of mathematical sciences research for the rewards structure indicate that "research in the discipline" was almost universally seen as very important and that it should be very important. Results also indicated that "interdisciplinary research involving new mathematics" and "applications of existing mathematics to other fields" were seen as important, but not as important as "research in the discipline".

4. There is ambiguity and uncertainty in the mathematical sciences community about what should be included in the definition of scholarship.

5. Lack of effective communication between various organizational levels is a major problem at many institutions.

6. A. The role of the chair is critical to the well-being of the department.
   B. There are marked discrepancies between the answers of the chairs and faculty on many questions in the survey.

7. There is general dissatisfaction with the methods of evaluating teaching, especially student evaluation questionnaires on teaching.

8. There is discomfort with the evaluation of faculty duties in general.

9. "Quality-of-life" issues are of major importance in any rewards structure.

10. Most faculty members favor a rewards system that includes a combination of across-the-board and merit increases.

Guiding Principles

1. Research in the mathematical sciences and its applications is fundamental to the existence and utility of the discipline and should continue to be among the primary factors of importance in the recognition and rewards systems.
2. Each department should ensure that contributions to teaching and related activities and to service are among the primary factors of importance in the recognition and rewards system.
3. Departments should develop policies that encourage faculty to allocate their efforts in ways that are as consistent as possible with their current interests and, at the same time, fit the needs of the department. The goal should be to create a department that meets all its obligations and aspirations with excellence, while at the same time engaging faculty in activities that they find personally rewarding. These activities should be recognized as valuable, and they should be rewarded when done well.

Talking about Leaving

- *Factors Contributing to High Attrition Rates among Science, Mathematics & Engineering Majors, Elaine Seymour and Nancy M. Hewitt, Bureau of Sociological Research, Westview Press, Boulder, CO, 1994, 1997*

Within two years after taking a college science, mathematics, or engineering class, 40%–60% of a group of above-average students have left majors in these disciplines. This report endeavors to document reasons that had been given with smaller, earlier studies. In the 1980s Treisman and Henkin wondered why so few African-American students succeeded in introductory calculus at Berkeley. They wrote a list of reasons they guessed for why students would not succeed. Their list was not very different from the one given in this book. Faculty believe the reasons students leave include:

- Some students choose the wrong area to begin with.
- Some students are underprepared.
- Some students lack interest, ability, competence, or capacity for hard work.
- Some students discover a passion for another discipline.

Treisman and Henkin discovered that the data at Berkeley did not support these reasons. Part of Treisman's work was to isolate more significant factors. This ethnographic study underscores the factors that seem to play a significant role. Factors that are part of the educational experience and the culture in science, mathematics, and engineering seem to be most significant. Students who leave and those who stay in these disciplines repeatedly mention the same factors.

Hewitt and Seymour describe two groups of students as "more pulled than pushed" and "more pushed than pulled" away from studying science, mathematics, or engineering. The students in the first group are often ambivalent about switching and may feel they will someday return. "They attribute their decision to leave almost exclusively to the poverty of the educational experience created by the weed-out system, and, by any measure, represent a loss to science, mathematics, and engineering of high-quality students."

The second group have the ability, are adequately prepared, and entered majors with interest. Poor teaching and a weed-out environment discourages these students. They enter other majors that they view as a poor compromise. These students are frequently angry, resentful, regretful, and frustrated because they feel science, mathematics, or engineering is the right choice for them. They believe they could succeed given the right support and a less competitive atmosphere. Many females and students of color fall into this group.

Seymour and Hewitt did encounter students leaving for the reasons stated at the outset. However, they hypothesize that "on every campus, there are substantial numbers of students who could be retained in S.M.E. majors if appropriate structural and cultural changes are made." Some of the case studies cited in the references support this hypothesis, at least for mathematics.

Seven different institutions of varying type and about 460 students participated in this study. The hypotheses of this study are not original, and the authors have thoroughly investigated, analyzed, and reported on earlier studies with smaller numbers of students.

The authors look at differences among institutions; choice of major; preparation for college study; difficulty of science, mathematics, and engineering majors; the competitive environment of these majors; the teaching and learning environment; issues of career, money, time, and lifestyle; gender issues; and issues of ethnicity. Mingled with statistics about student motivations are a very large number of quotations by students that accord with the statistics.

The SIAM Report on Mathematics in Industry

- *Report prepared by the Society of Industrial and Applied Mathematics, Philadelphia, PA, 1995*

This can be found at http://www.siam.org/mii/index.htm.

The report presents the results of a survey of nonacademic mathematicians and their managers about the mathematics they use, the problems to which their mathematics is applied, the environment in which they work, and their assessment of the strengths and weaknesses of graduate training in mathematics. The report ends with suggestions for making graduate training in mathematics more responsive to the needs of nonacademic mathematicians. The study presents a substantial set of useful survey data.

The Role of Mathematics in Industry

Employment of Nonacademic Mathematicians by Degree and Field

	Ph.D.	*M.S.*
Government	28%	22%
Engineering research, computer software and services	19%	18%
Manufacturing (electronic, computers, aerospace, transportation)	17%	12%
Services (financial, communications, transportation)	13%	22%
Chemical, pharmaceutical, petroleum	6%	2%

The study found that many areas of pure mathematics as well as most all areas of applied mathematics found use in industry and government. For example, algebra and number theory were used in cryptography, formal systems and logic were used in computer security and verification, and geometry was used in computer-aided engineering and design. Nearly every manager interviewed cited particular problems where mathematics had made a significant contribution. The mathematical reasoning skills cited by managers as of greatest value were: modeling and simulation, mathematical formulation of problems, algorithm and software development, problem solving, statistical analysis, verifying correctness, and analysis of accuracy and reliability.

Both managers and mathematicians indicated that they saw substantial new opportunities for mathematicians in industry and government. Manufacturing, product development, and materials were listed as particularly promising areas.

The Working Environment

Some of the key findings about the role of mathematicians and the R&D context in which they work are:

- Mathematicians are part of the R&D infrastructure; mathematics cannot be viewed as an end in itself.
- Nonacademic research is often faulted for too much understanding with too little transfer. Even in groups with a research charter, examples of success with products or services are required to justify continued support.
- Mathematicians are typically scattered across an organization among engineers, physicists, and computer scientists, where they are supported by various mission-oriented groups.

- Nonacademic mathematicians need to be facile at working with a wide range of mathematical skills in support of projects. Even a single project will have many aspects requiring a variety of mathematical techniques. At the same time, it is desirable to have special expertise in some area. Further, nonacademic mathematicians need to have an interest and some knowledge in other technical areas. This is important for developing real solutions to real-world problems. The most frequent discipline cited was computer science.

Formulating problems was found to be an interactive and continuing process for mathematicians working on projects with other R&D scientific staff. Good communicating and listening skills, as well as general interpersonal skills, are critical. The hardest task for a mathematician is typically developing the real-problem requirements. The user does not usually know what the solution will look like in the end. Mathematicians cannot throw their solutions "over the wall" and be done with a project. Customers inside or outside one's organization may express frustration with the current solution without communicating clearly what they really want. Indeed, a mathematician's biggest contribution to a team is often the ability to pose the right question. In addition, nonacademic mathematicians can be expected to provide a "solution" even when no rigorous solution can be found or when there is not time to find one.

Interviews consistently found that mathematicians are valued most of all for two general attributes: highly developed skills in abstraction, analysis of underlying structures, and logical thinking; and expertise with the best tools for formulation and solving problems.

Well-trained, even pure, mathematicians were viewed as critically equipped to keep going when textbooks have to be left behind. Mathematicians are seen as better equipped than others in coming up with the correct definitions of problems and developing the right level of abstraction. Mathematicians were also cited for their ability to spot hidden gaps in the analysis of a problem and to identify connections.

Shortcomings of some mathematicians who did not fully understand the nature of the nonacademic environment were, according to managers: a tunnel vision (writing a paper and that's the solution); a lack of concern for the real environment that requires realistic models, cost considerations, and implementation details; and the desire to continue investigations forever instead of recognizing when to stop.

Perceptions of Graduate Education

While a number of reports have voiced the concern that graduate education is only training students to be the clones of their professors, nonacademic mathematicians interviewed in this study mostly believed that their graduate education had helped them to obtain and perform well in their present positions. They felt that their graduate education had been very effective in developing facility in:

- logical thinking and the ability to deal with complexity,

- broadly applicable problem solving,
- conceptualizing and abstracting,
- formulating problems and modeling.

Many nonacademic mathematicians felt that their graduate preparation was wanting in aspects outside their core mathematical training. The areas where preparation was rated as less than good included:

- working well with colleagues,
- communicating at different levels,
- having broad scientific knowledge,
- effectively using computer software.

These problems were substantial enough that 90 percent of Ph.D.'s and M.S.'s interviewed said that it was important to make educational changes in graduate mathematics training.

Managers echoed the problem areas of their mathematical employees, saying that they felt improvement was needed in graduate mathematics training in the areas of applications of mathematics, knowledge of other disciplines, real-world problem solving, oral and written communication, computer skills, and teamwork.

Suggestions and Strategies

The suggestions in this study for changes in graduate mathematics education largely mirror the areas that nonacademic mathematicians and their employers cited in the previous section as in need of improvement: substantive exposure to applications of mathematics in the sciences and engineering; experience in formulation and solving real-world problems, preferably involving a variety of disciplines; computation; and communication and teamwork. For faculty the report recommends activities to enhance connections between mathematics faculty and researchers in other disciplines, inside and outside academia. For graduate students there are recommendations for taking the initiative in making contact with nonacademic mathematicians and researchers in other disciplines. For nonacademic organizations that use mathematicians there are recommendations for building various connections with university mathematicians and their students.

Shaping the Future—New Expectations for Undergraduate Education in Science, Mathematics, Engineering, and Technology

- *Advisory Committee to the National Science Foundation, Directorate for Education and Human Resources, National Science Foundation, Washington, DC, 1996*

In 1996 the National Science Foundation released "Shaping the Future", a report of the Advisory Committee to the Directorate for Education and Human Resources. The report was created by the Advisory Committee's Subcommittee on Undergraduate Education, under the leadership of Dr. Melvin D. George. Dr. George is a mathematician and a retired president of both St. Olaf College and the University of Missouri.

Dr. George's committee was charged with conducting an intensive review of the state of undergraduate education in science, mathematics, engineering, and technology in America and to prepare a report that was "action oriented, recommending ways to improve undergraduate education in science, mathematics, engineering, and technology."

The EHR Advisory Committee unanimously approved and endorsed the report. Since the release of the report, the NSF has co-sponsored a large number of "Shaping the Future" conferences at colleges and universities across the country. By their actions the EHR is demonstrating their strong support of the report. It is reasonable to assume that funding decisions made by EHR in coming years will be designed to further support the recommendations in the report.

The "Shaping the Future" report focused on recommendations to support one major goal:

> All students should have access to supportive, excellent undergraduate education in science, mathematics, engineering, and technology, and all students should learn these subjects by direct experience with the methods and processes of inquiry.

The report offered a variety of recommendations to institutions of higher education; business, industry, and the professional community; national and regional media; governments at the state and federal level; and the NSF. We reproduce here some of the recommendations that might most directly impact the professional lives of mathematicians in our colleges and universities.

- The president and the Congress: Establish, in consultation with the higher education community, a new social contract for higher education in America. What is needed may be a new act to reconnect the research base of these institutions to the learning of students and to service to the wider community.

- State governments: Ensure that funding formulas and state policies are modified, as necessary, to provide incentives and rewards for increased undergraduate student learning in science, mathematics, engineering, and technology (SME&T) at institutions in the state.

- University administrators:

  1. Reexamine institutional missions in light of needs in undergraduate SME&T-education.

  2. Hold accountable and develop reward systems for departments and programs, not just individuals, so that the entire group feels responsible for effective SME&T learning for all students.

3. Create or strengthen an institution-wide commitment to the preparation of K–12 teachers and principals, bringing together departments of education, SME&T and other departments, K–12 staff, and employers of teachers to design and implement improved teacher preparation programs having substantial SME&T content and stressing rigorous standards, along with emphasis on engaging students in learning.

- Departments:
  1. Encourage faculty to work toward the understanding of and resolution of serious educational issues, and reward those who most effectively help all students learn.
  2. Provide opportunities for graduate students to learn about effective teaching strategies as part of their graduate program.
- Professional societies: Work together to promote education as well as research, focus on student learning as well as teaching, and help departments in their disciplines find realistic ways to implement these recommendations.
- NSF:
  1. Lead the development of a common national agenda for improving undergraduate SME&T education in a collaborative way with other Federal agencies and foundations.
  2. Make clear to all colleges, universities, and other educational institutions receiving grants and contracts that the NSF expects its awards to contribute positively to the quality of undergraduate SME&T education.

In total, the Shaping the Future Report offers nearly 100 recommendations to every possible participant in the business of educating undergraduates in science, mathematics, engineering, and technology. Given the high visibility that NSF is giving to this report, mathematics departments are well advised to believe that the recommendations of this report will drive funding decisions at the Foundation.

Reinventing Undergraduate Education: A Blueprint for America's Research Universities

- *Boyer Commission for Educating Undergraduates at the Research University, Carnegie Foundation, Stony Brook, New York, 1998*

This can be found on the Web at: http://notes.cc.sunysb.edu/Pres/boyer.nsf.

Background: This is a publication of the Boyer Commission on Educating Undergraduates in the Research University, which was funded by the Carnegie Foundation for the Advancement of Teaching. The Commission was named for

Ernest Boyer, president of the Carnegie Foundation until his death in 1995. Bruce Alberts, president of the NAS, was one of eleven members of the Commission.

This report is a fairly broad-sided attack on the quality of undergraduate education at America's research universities. While the 125 universities that are classified as Research I or II institutions comprise only 3% of the 3,500 institutions of higher education, they award 32% of the undergraduate degrees in America and 56% of the baccalaureates in science and engineering during the period 1991–95. In all science fields except chemistry, the majority of students who obtain a Ph.D. earned their bachelor's degree at a U.S. research university.

Mathematics is almost invisible in the report. It is mentioned only a couple of times and then in connection with remedial education or the teaching of freshmen by graduate students.

Basically, the report is a call for dramatic changes in how research universities teach undergraduates. The goal is to create an undergraduate experience (centered on inquiry learning and research experiences for undergraduates) that is not duplicated by the other types of institutions that award undergraduate degrees. The centerpiece of the report is an Academic Bill of Rights and a set of ten guiding principles for changing undergraduate education. Each principle is followed by a set of recommendations to implement the principle.

The Academic Bill of Rights asserts that by admitting a student, a college or university should commit to providing the following:

At all colleges and universities:

- Opportunity to learn through inquiry;
- Training in the skills necessary for oral and written communication;
- Appreciation of arts, humanities, sciences, and social sciences;
- Careful and comprehensive preparation for whatever may lie beyond graduation.

Additional rights for students at research universities:

- Expectation of and opportunity for work with talented senior researchers;
- Access to first-class facilities in which to pursue research;
- Many options among fields of study;
- Opportunities to interact with people of backgrounds, cultures, and experiences different from the student's own.

The ten guiding principles are:

1. Make research-based learning the standard.
2. Construct an inquiry-based freshman year.
3. Build on the freshman foundation.
4. Remove barriers to interdisciplinary education.
5. Link communication skills and course work.

6. Use information technology creatively.
7. Culminate with a capstone experience.
8. Educate graduate students as apprentice teachers.
9. Change faculty reward systems.
10. Cultivate a sense of community.

At present this report has received much criticism from the academic community. Indeed, one is led to conclude that it is unlikely that it will have as much impact as did "Scholarship Reconsidered" (an earlier report by the Carnegie Foundation in 1990; see p. 232.)

Report of the Senior Assessment Panel of the International Assessment of the U.S. Mathematical Sciences

- *A report commissioned by the National Science Foundation using a panel of mathematicians drawn largely from outside the United States as well as scientists from related disciplines, 1998.*

This report was prepared for the National Science Foundation (NSF) by a panel of individuals who had not received funding from the Foundation. It was prepared in response to the Government Performance and Results Act, which called for agencies to set strategic goals and evaluate their progress toward those goals. The panel was charged with making specific recommendations to the Foundation. A brief summary of the report can be found in the *Notices of the AMS* **45**, no. 7 (1998), 880-82. The report also contains the article by Gromov, which is included as Chapter 19 of this book. Appendix 2 of the report contains the panel's assessment of the health of various subdisciplines of mathematics in the United States.

The report points out the disparity between the percentage of scientists with federal support as a fraction of scientists active in research in the various disciplines: 69% in biological sciences, 67% in physical sciences, and 35% in mathematics.

The panel notes the ways in which mathematics research differs from research in the other sciences. It is small science, and much work is done by individuals working alone, with modest equipment needs. Mathematical research is long-lasting, which forces a need for good libraries. Mathematics is an international discipline; thus local events can lead to widespread migration of mathematicians, as from Europe before World War II or from Eastern Europe more recently.

The percent of Ph.D.'s going to noncitizens is 55% in the U.S., 33% in France, 40% in Japan, and 27% in England.

The panel did its benchmarking by considering the contributions of the U.S. mathematics community to fundamental mathematics by assessing interactions between mathematics and the users of mathematics and by assessing the quality of undergraduate, graduate, and postdoctoral education.

The comparisons involve data (the number of research papers by regions of the world, the number of speakers at the International Congress by region) which are not perfect (for example, there is no good measure of the number of Ph.D.'s in the regions).

The panel felt that communication between mathematical scientists and other scientists is poor the world over but that several countries were becoming more involved in promoting multidisciplinary research. They felt that the U.S. undergraduate programs offer less exposure to mathematics than programs in Europe and Asia.

U.S. graduate programs offer a wider range of specialization, but European programs offer better financial support to graduate students. Retention in U.S. graduate programs is lower than in Europe (particularly for U.S. students). Graduates of U.S. doctoral programs have higher expectations of an academic career than in Europe, while academic jobs are constant or decreasing. All of these observations may play a role in the fact that the percentage of U.S. students pursuing graduate degrees has declined in recent years (although a loss of interest in graduate work in mathematics has been seen in several nations).

The report contains an analysis of sources of federal support showing mathematics to be more heavily dependent on NSF funding (60% of 1997 federal funding for mathematics), to be more dependent on institutional support for graduate students, and to have only a small share of overall federal support for academic basic research. Nonfederal support is extremely limited.

The report contains the following overall assessment: The U.S. has the lead in many subdisciplines and is capable of responding to breakthroughs in all areas of mathematics. Yet U.S. mathematics suffers from isolation from the rest of science, a decline in the number of young people entering the field, and a low level of interaction with nonacademic fields, particularly in the private sector. The panel concludes that morale is low in the U.S. The European Union is expanding opportunities and funding for young mathematicians, while in the U.S. students are overly dependent on teaching, which extends their time to degree and decreases the attractiveness of mathematics to young people.

Specific findings of the panel are:

- **Finding 1:** The academic success of U.S. mathematics has been and remains distinguished. Although the U.S. is the strongest national community in the mathematical sciences, this strength is somewhat fragile. U.S. strength rests heavily on mathematicians who have come from outside the U.S. The lack of financial support thwarts the careers of many young mathematical scientists.
- **Finding 2:** Academic mathematics is insufficiently connected to mathematics outside the university. (The report makes it clear that each side could do a better job of reaching out to the other side.) Academic mathematics could interact fruitfully with other disciplines in ways that are often obscured by the inward focus of mathematics and science departments. The structure of universities mitigates against multidisciplin-

ary research. Scientific problems of the future will be extremely complex and will require collaborative mathematical modeling, simulation, and visualization. (The panel urges funding agencies to provide financial support that recognizes and rewards multidisciplinary activities and that recognizes the long time required to become competent in such work.)

- **Finding 3:** U.S. graduate programs in the mathematical sciences, especially the top 25, are considered to be among the very best in the world. Graduate applications in the mathematical sciences have declined, however. Careers in mathematics have become less attractive to U.S. students. The curriculum in U.S. institutions for undergraduates needs to be strengthened, broadened, and designed for more active participation by students in discovery. There are exciting mathematical science career opportunities outside the academy.

The panel recommends to the mathematical sciences community:

- Academic mathematical science must strike a better balance between theory and application.
- For U.S. mathematical sciences to thrive, the discipline must be made more attractive to young Americans with bright and inquisitive minds.

It recommends to NSF:

- NSF's specific objective should be to build and maintain an academic community in mathematics that is intellectually distinguished and relevant to society.
- NSF's broad objectives should be to build and maintain the mathematical sciences in the U.S. at the leading edge of the mathematical sciences and to strongly encourage it to be an active and effective collaborator with other disciplines and with industry.

The panel suggests some strategies to accomplish these objectives:

- Bring the number of active researchers supported to a level comparable to those in the physical and biological sciences and engineering.
- Encourage activities that connect mathematics to areas of application.
- Strengthen the connection between pure and applied mathematics.
- Broaden the exposure of mathematicians to problems in other fields.
- Maintain and strengthen abstract mathematics.

Unlocking Our Future: Toward a New National Science Policy

- *A report to Congress by the House Committee on Science, 1998. (The subcommittee writing the report was chaired by Congressman Vern Ehlers, who holds a Ph.D. in physics.)*

This report can be found at
http://www.house.gov/science/science_policy_study.htm.

Overview

The growth of economies throughout the world since the industrial revolution began has been driven by continual technological innovation through the pursuit of scientific understanding and application of engineering solutions. America has been particularly successful in capturing the benefits of the scientific and engineering enterprise, but it will take continued investment in this enterprise if we hope to stay ahead of our economic competitors in the rest of the world. Many of those challengers have learned well the lessons of our employment of the research and technology enterprise for economic gain.

Americans must remain optimistic about the ability of science and engineering to help solve their problems and about their own ability to control the application of technological solutions. The United States of America must maintain and improve its preeminent position in science and technology in order to advance human understanding of the universe and all it contains and to improve the lives, health, and freedom of all peoples. The continued health of the scientific enterprise is a central component in reaching this vision. In this report, therefore, we have laid out our recommendations for keeping the enterprise sound and strengthening it further. There is no singular, sweeping plan for doing so. The fact that keeping the enterprise healthy requires numerous actions and multiple steps is indicative of the complexity of the enterprise. The fact that this report advocates not a major overhaul but rather a fine-tuning and rejuvenation is indicative of its present strength.

This report focuses on three major areas: (1) government's role in supporting the research enterprise; (2) the private sector's role in supporting the research enterprise; and (3) the collective responsibility of government, industry, and educators to strengthen science and mathematics education. In addition, the report discusses the need for science to play a greater role in public policy and international relations and the need for science to reforge its ties with the American people to gain their support and trust.

Recommendations of Interest to Mathematicians

Importance of Basic Research. It is in the country's interest for its scientists to continue pursuing fundamental, ground-breaking research. The experience with fifty years of government investment in basic research has demonstrated the economic benefits of this investment. To maintain the nation's economic strength and international competitiveness, Congress should provide stable and substantial federal funding for scientific research.

Basic Research Is a Federal Special Priority. Fiscal reality requires setting priorities for spending on science and engineering. Because the federal government has an irreplaceable role in funding basic research, priority for federal funding should be placed on fundamental research. Moreover, because innovation and creativity are essential to basic research, the federal government should consider allocating a certain fraction of grant monies specifically for creative, ground-breaking research.

Breadth of Federal Support to Basic Research. The practice of science is becoming increasingly interdisciplinary, and scientific progress in one discipline is often propelled by advances in other, seemingly unrelated fields. It is important that the federal government fund basic research in a broad spectrum of scientific disciplines, mathematics, and engineering and resist concentrating funds in a particular area.

Limited Role for Government in Applied Research. While the federal government may, in certain circumstances, fund applied research, there is a risk that using federal funds to bridge the mid-level research gap could lead to unwarranted market interventions and less funding for basic research. It is important, therefore, for companies to realize the contribution investments in mid-level research can make to their competitiveness. The private sector must recognize and take responsibility for the performance of research.

Partnerships in Research. Partnerships in the research enterprise can be a valuable means of getting the most out of the federal government's investment. Partnerships between university researchers and industries have become more prevalent, and should be encouraged, as a way for universities to leverage federal money and for industries to capture research results without building up in-house expertise. However, the independence of the institutions and their different missions need to be respected. International scientific collaborations form another important aspect of the research enterprise and are often essential for large-scale scientific projects like the international space station.

Partnerships for Economics Development. Partnerships that tie together the efforts of state governments, industries, and academia also show great promise in stimulating research and economic development. Indeed, states appear far better suited than the federal government to foster economic development through technology-based industry.

Partnerships for Strengthening Regional Research. The university community has a role in improving research capabilities throughout its ranks, especially in states or regions trying to attract more federal R&D funding and high-tech industries. Major research universities should cultivate working relationships with less well-established research universities and technical colleges in research areas where there is mutual interest and expertise, and consider submitting, where appropriate, joint grant proposals. Less research-intensive colleges and universities should consider developing scientific or technological expertise in niche areas that complement local expertise and contribute to local economic development strategies.

Scientific Partnerships with Policymakers. For science to play any real role in legal and policy decisions, the scientists performing the research need to be seen as honest brokers. To ensure that decision makers are getting sound analysis, all federal government agencies pursuing scientific research, particularly regulatory agencies, should develop and use standardized peer review procedures. In return, scientists should be required to divulge their credentials, provide a résumé, and indicate their funding sources and affiliations when formally offering expert advice to decision makers. In Congress and the executive branch, science policy and funding remain scattered piecemeal over a broad range of committees and departments. These diffusive arrangements make effective oversight and timely decision making extremely difficult.

The Critical Importance of Outstanding Science and Mathematics Education. No factor is more important in maintaining a sound R&D enterprise than education. Yet student performance on the recent Third International Math and Science Study highlights the shortcomings of current K–12 science and math education in the U.S. We must expect more from our nation's educators and students if we are to build on the accomplishments of previous generations. New modes of teaching math and science are required. Curricula for all elementary and secondary years that are rigorous in content, emphasize the mastery of fundamental scientific and mathematical concepts as well as the modes of scientific inquiry, and encourage the natural curiosity of children must be developed.

Attracting Qualified Teachers. It is necessary that a sufficient quantity of teachers well versed in math and science be available. Programs that encourage recruitment of qualified math and science teachers, such as flexible credential programs, must be encouraged. In general, future math and science teachers should be expected to have had at least one college course in the type of science or math they teach, and preferably at least a minor. Ongoing professional development for existing teachers also is important. Another disincentive to entry into the teaching profession for those with a technical degree is the relatively low salaries K–12 teaching jobs offer compared to alternative opportunities. To attract qualified science and math teachers, salaries that make the profession competitive may need to be offered. School districts should consider merit pay or other incentives as a way to reward and retain good K–12 science and math teachers.

Opportunities in Educational Technology. The revolution in information technology has brought with it exciting opportunities for innovative advances in education and learning. As promising as these new technologies are, however, their haphazard application has the potential to adversely affect learning. A greater fraction of the federal government's spending on education should be spent on research programs aimed at improving curricula and increasing the effectiveness of science and math teaching.

Challenges in Graduate Education. Graduate education in the sciences and engineering must strike a careful balance between continuing to produce the world's premier scientists and engineers and offering enough flexibility so that students with other ambitions are not discouraged from embarking on further education in math, science, or engineering. While continuing to train scientists and engineers of unsurpassed quality, higher education should also prepare students who plan to seek careers outside of academia by increasing flexibility in graduate training programs. Specifically, Ph.D. programs should allow students to pursue course work and gain relevant experience outside their specific area of research.

The length of time involved and the commensurate forfeiture of income and benefits in graduate training in the sciences and engineering is a clear disincentive to students deciding between graduate training in the sciences and other options. Universities should be encouraged to put controls on the length of time spent in graduate school and postdoctoral study and to recognize that they cannot attract talented young people without providing adequate compensation and benefits.

Increased support for master's programs is needed to allow students to pursue an interest in science without making the long commitment to obtaining a Ph.D., thus attracting greater numbers of students to careers in science and technology. More university science programs should institute specially designed master of science degree programs as an option for allowing graduate study that does not entail a commitment to the Ph.D.

Importance of Postdoctoral Training. The training of scientists and engineers in the U.S. occurs largely through an apprenticeship model, in which a student learns how to perform research through hands-on experience under the guidance of a thesis advisor. A result of this link between education and research is that students and postdoctoral researchers are responsible for actually performing much of the federally funded research done in universities. Mechanisms for direct federal funding of postdocs are already relatively common. Expansion of these programs to include greater numbers of graduate students in math, science, and engineering should be explored.

Communication Problems. Educating the general public about the benefits and grandeur of science is also needed to promote an informed citizenry and maintain support for science. Both journalists and scientists have responsibilities in communicating the achievements of science. However, the evidence suggests that the gap between scientists and journalists is wide and may be getting wider. Closing it will require that scientists and journalists gain a greater appreciation for how the other operates.

As important as bridging the gap between scientists and the media is, there is no substitute for scientists speaking directly to people about their work. In part because science must compete for discretionary funding with disparate interests, engaging the public's interest in science through direct interaction is crucial. Scientists and engineers should be encouraged to take time away from their research to educate the public about the nature and importance of their work.

Chapter 23
Where to Find Other Material

Reports about Mathematics Research

Renewing U.S. Mathematics, National Research Council, National Academy Press, 1984.

This is the famous "David Report", which documented exciting developments in the mathematical sciences and effectively made the case for increased NSF funding in the mathematical sciences. (Note that David was an eminent engineer from industry, not an academic mathematician.) For a full discussion, see Chapter 22.

Renewing U.S. Mathematics: A Plan for the 1990's, National Research Council, National Academy Press, 1991.

An update of the "David Report". Along with urging the government and university administrations to make more money available for mathematics, the report presents some challenges involving better career paths for young mathematicians and balancing teaching with research for faculty and graduate TAs. A broadened training at the doctoral level is also urged.

Report on the Senior Assessment Panel of the International Assessment of the U.S. Mathematical Sciences, National Science Foundation, 1998.

This Congressionally mandated "benchmarking" assessment of U.S. mathematics also contains an honest assessment of the strengths of the American mathematics research enterprise, both overall and in a field-by-field breakdown. The report also contains important recommendations for how NSF should support mathematics.

Mathematical Sciences, Technology and Economic Competitiveness, J. Glimm, ed., National Research Council, 1991.

This document provides a good foundation in industrial mathematics from the point of view of clients, presenting the priorities of industry and federal agencies. The report gives an overview of mathematical sciences-based technology transfer to the business and governmental sectors as it summarizes the opportunities and challenges.

SIAM Report on Mathematics in Industry, Society for Industrial and Applied Mathematics,1995.
This study examines the roles of mathematics outside academia as well as the skills and preparation needed by nonacademic mathematicians. The report suggests strategies for enhancing graduate education in mathematics and nonacademic career opportunities for mathematicians. For a full discussion, see Chapter 22.

Preserving Strength While Meeting Challenges, Board of Mathematical Sciences, National Science Council, 1997.
Proceedings of a BMS workshop with papers about how the public views science and mathematics, how scientists view the role of mathematicians, the challenges to NSF, the challenges in the education arena, and areas of new opportunities for the mathematical sciences.

General Reports about Research and Education in Universities

Renewing the Promise: Research-Intensive Universities and the Nation, President's Council of Advisors on Science and Technology, 1992.
Among the findings and recommendations are: challenges in adapting to a tighter resource environment; better collaboration between academia and industry; developing a better balance between research, teaching, and outreach; and developing an honest strategic-planning process. Universities and individual departments are encouraged to develop focused strengths.

In the National Interest: The Federal Government and Research-Intensive Universities, President's Council of Advisors on Science and Technology, 1995.
This report reinforces many of the findings in "Renewing the Promise: Research-Intensive Universities and the Nation" above. Among other recommendations, it calls for a change in the reward system to encourage senior faculty to be more involved in undergraduate teaching and student advisement; more outreach by research universities to two-year colleges, and more attention to training of pre-service school science and mathematics teachers.

Reinventing Undergraduate Education: A Blueprint for America's Research Universities, Carnegie Foundation, Stony Brook, New York, 1998.
This report by the Boyer Commission for Educating Undergraduates at Research Universities is quoted in Chapter 1 of this book. The report has been controversial among academics for its opening criticism of how research universities are not giving undergraduates the attention they deserve. However, it goes on to give thoughtful advice on how research universities can exploit their strengths to infuse the spirit of research into undergraduate education. For a full discussion, see Chapter 22.

Beginning a Dialogue on the Changing Environment for the Physical and Mathematical Sciences, National Research Council, 1994.
The proceedings of an NRC workshop concerned with the need to effect significant changes in both the research and educational missions of universities. While much of this report focuses on concerns of laboratory sciences, it has some rec-

ommendations relevant to mathematicians. These include: (1) instead of focusing on resources, we need to think about how research can develop and thrive in a changing environment; and (2) institutional reform will probably be required if we are to get the most science possible out of the resources that are available.

Reports about Doctoral and Postdoctoral Training in Mathematics

Educating Mathematical Scientists: Doctoral Study and the Postdoctoral Experience in the United States, Ron Douglas, ed., Board of Mathematical Sciences, National Research Council, 1992.

This document thoughtfully dissects graduate training into a number of components and analyzes critical issues in each. The report summarizes practices of a number of successful doctoral programs in mathematics. It presents three models for a mathematics department. The most relevant one for most departments is the specialized model, in which a department has half or more of its faculty in one area for a focused strength. The report anticipates several subsequent NRC reports in arguing for broader doctoral and postdoctoral training to prepare Ph.D.'s for a variety of nonacademic jobs. There are helpful suggestions about general issues, like recruiting and retaining doctoral students, and specific issues, like placing foreign students with weak language skills but advanced training in mathematics. For a full discussion, see Chapter 22.

Graduate Education and Postdoctoral Training in the Mathematical Sciences, National Science Foundation, 1996.

Findings and recommendations emerging from this 1995 NSF workshop address: broadening the intellectual content and increasing the diversity of skills acquired during Ph.D. training, adjusting the balance between research and education in doctoral and postdoctoral training, shortening the time to completion of the Ph.D., increasing internships and other real-world experiences, and changing graduate student support mechanisms in the NSF.

Report about Mathematics Education at All Levels

Everybody Counts, Mathematical Sciences Education Board, National Research Council, 1989.

This famous document is a case statement for the growing importance of a strong mathematical education for all students. While aimed primarily at precollege instruction, the report provides excellent support for collegiate mathematics.

Reports about Undergraduate Mathematics Education and Its Recognition

Recognition and Rewards in the Mathematical Sciences, Joint Policy Board for Mathematics, American Mathematical Society, 1994.

The report presents an in-depth study involving site visits to twenty-six mathematics departments. An important observation is that faculty members' contributions to their departments are likely to change over their lifetimes. A key finding was that faculty and department chairs have different perceptions of the reward structure, with chairs believing that teaching carries more weight than faculty believe it carries. Communication problems and the importance of a strong department chair are other critical issues discussed. The report recommends that each department should develop a working definition of scholarship that is consistent with the departmental and institutional missions and is sufficiently encompassing and flexible to embrace the broad variety of intellectual activities in the discipline. For a full discussion, see Chapter 22.

Moving Beyond Myths, Mathematical Sciences Education Board, National Research Council, 1991.

This report's main recommendations are echoed in later reports; they include greater recognition for teaching and more involvement of university mathematicians in school mathematics. The report was vocally criticized in the research mathematics community for its negative assessment of current instructional practices, such as many faculty's alleged weak commitment to teaching. However, it is instructive to read this report as a reflection of troublesome perceptions—heard again in this Task Force's focus groups with deans (see Chapter 6)—about mathematicians that exist in campus administrations, across the sciences and outside of academia. For a full discussion, see Chapter 22.

Models That Work: Case Studies in Effective Undergraduate Mathematics Programs, A. Tucker, ed., MAA Notes #38, Mathematical Association of America, 1996.

This study, cited in Chapter 13, discusses common themes of effective undergraduate mathematics programs. Most of the report's findings build on site visits to ten successful mathematics programs, ranging from a two-year college to two Top 10 research universities. One theme is the encouragement of continual experimentation in individual faculty classes.

Assessing Calculus Reform Efforts, J. Leitzel and A. Tucker, Mathematical Association of America, 1995.

This report summarizes the impact of the NSF Calculus Reform Initiative up through 1994. It documents surprisingly widespread experimentation with calculus reform at research universities. A key finding is how satisfied most faculty teaching reformed courses were with the levels of interest and performance of their students.

Guidelines for Programs and Departments in Undergraduate Mathematics, Mathematical Association of America, 1993.

While this report is aimed primarily at the assessment of four-year college mathematics departments, it conveniently inventories many concerns that are applicable to university mathematics departments.

Recommendations for a General Mathematical Sciences Program, Mathematical Association of America, 1980. Reprinted in Reshaping College Mathematics, MAA Notes #13, 1989.

The last comprehensive CUPM report on the undergraduate major in mathematics. While almost twenty years old, this report presents an inclusive view of the mathematics major that is getting more and more currency today.

Mathematics Outside of Mathematics Departments: A Study of Mathematics Enrollment's in Non-Mathematics Departments, by S. Garfunkel and G. Young, Consortium for Mathematics and Its Applications, 1990 (summarized in the November 1990 AMS Notices).
This study documents that more upper-division mathematics instruction, primarily in applied mathematics, occurs outside mathematics departments than inside them.

Challenges for College Mathematics: An Agenda for the Next Decade, L. Steen, ed., Report of the American Association of Colleges, reprinted in FOCUS, November 1990, pp. 1–32.
A report focusing largely on noncurricular components of undergraduate mathematics instruction, such as hurdles and strategies for making mathematics classrooms more welcoming to students from underrepresented groups.

Crossroads in Mathematics: Standards for Introductory College Mathematics before Calculus, American Mathematical Association of Two-Year Colleges, 1995.
A companion document for two-year colleges to the "MAA Guidelines for Programs and Departments in Undergraduate Mathematics" above.

Twenty Questions that Deans Should Ask Their Mathematics Departments, by L. Steen, Bulletin of the American Association of Higher Education, May 1992.
While not a report, this article is similar to some of the preceding reports in its underlying concerns. It anticipates many of the findings and recommendations of this resource book. Along with the questions are principles to aid mathematics departments to be ready for the questions.

Reports of NSF Undergraduate Education Self-Assessments

Shaping the Future: New Expectations for Undergraduate Education in Science, Mathematics, Engineering, and Technology, National Science Foundation, 1996.
This review of the NSF's Division of Undergraduate Education makes recommendations to the NSF for future funding priorities in undergraduate education. There are also thoughtful suggestions to universities and faculty that rise above specific knowledge acquisition to address issues such as lifelong learning, critical thinking, and communication skills. For a full discussion, see Chapter 22.

Evaluation of the Division of Undergraduate Education's Course and Curriculum Development Program, National Science Foundation, 1997.
This external evaluation of the main NSF undergraduate education program (which has since been reorganized) points to what works and what are critical factors for success in undergraduate instructional projects. It also suggests to the NSF future directions for funding instructional innovation and dissemination.

Resources for Improving Mathematics Teaching

Report of the Task Force on Teaching Growth and Effectiveness, Mathematical Sciences Education Board, National Research Council, 1993.

This report is a companion to "Recognition and Rewards in the Mathematical Sciences", cited above. It gives thoughtful guidance for faculty wading into the "mare's nest" of assessing effective teaching. A major theme is that effective teaching and growth in teaching are a collective responsibility of the faculty in a department. The report encourages senior faculty to assume a leadership role in the teaching arena by participating fully in teaching introductory courses, in course development, and in mentoring junior and part-time faculty and teaching assistants. An interesting finding was that NSF Presidential Young Investigators had complained about the low priority given to teaching (and public service) in their departments.

A Source Book for College Mathematics Teaching, A. Schoenfeld, ed., Mathematical Association of America, 1990.

This book contains a variety of information about teaching strategies. It goes into greater depth in analyzing ways to teach mathematics to undergraduates than any other publication.

How to Teach Mathematics, by S. Krantz, 2nd edition, American Mathematical Society, 1998.

A balanced, to-the-point paperback of helpful suggestions about the various issues associated with teaching mathematics in colleges and universities.

You're the Professor, What Next? Ideas and Resources for Preparing College Teachers, B. Case, ed., MAA Notes #35, Mathematical Association of America, 1994.

This resource book, prepared by the AMS/MAA Committee on Preparation for College Teaching, provides lots of good material about training graduate students to be future college teachers, including extensive descriptions of successful mathematics TA training programs at several universities. The report's appendices gather together in one place many short essays and reprints (from sources such as *UME Trends*) on teaching that are relevant for all faculty, as well as graduate TAs.

A Practical Guide to Cooperative Learning in Collegiate Mathematics, Nancy L. Hagelgans, ed., and Barbara Reynolds, MAA Notes 37, Mathematical Association of America, Washington, DC, 1995.

The handbook, prepared by the MAA's Advisory Board for Cooperative Learning in Undergraduate Mathematics Education, gives practical methods and a discussion of effectiveness of cooperative learning methods in college mathematics classrooms. There is helpful advice to first-timers about unanticipated difficulties.

McKeachie's Teaching Tips: Strategies, Research, and Theory for College and University Teachers, Wilbert J. McKeachie and Graham Gibbs, Houghton Mifflin, 1998.

This helpful book, which has gone through numerous editions, has advice about every aspect of collegiate teaching. Much of the book is about problems in non-quantitative courses, but its tips about generic issues, such as handling students who try to monopolize discussions and effective ways to discuss results of a test, make the book a valuable resource for new mathematics faculty.

Calculus: The Dynamics of Change, W. Roberts, ed., MAA Notes 39, Mathematical Association of America, 1996.

A summary of lessons learned in the ten years of calculus reform (since the 1986 Tulane Conference). This volume includes a helpful assessment of the resources required to undertake a major reworking of calculus instruction. There is also an article by Mort Brown describing the Michigan New Wave calculus in depth.

Resources for Calculus Collection: Volume 1: Learning by Discovery; Volume 2: Calculus Problems for a New Century; Volume 3: Applications of Calculus; Volume 4: Problems for Student Investigation, MAA Notes 27–30, Mathematical Association of America, 1994.

A helpful collection of calculus resource materials for enriching calculus instruction. The MAA Notes series has a number of additional calculus-enrichment volumes.

General Background on Undergraduate Education

How College Affects Students, E. Pascarella and P. Terenzini, Jossey-Bass, 1991.

This book presents findings and insights from twenty years of research on the subject. It attempts to distill a huge quantity of sometimes conflicting research.

Talking about Leaving: Why Undergraduates Leave the Sciences, Elaine Seymour and Gloria Hewitt, Westview Press, Boulder, CO, 1997.

A widely cited study interviewing students who did and did not drop out of science/math/engineering disciplines. Both groups complained about the way they were taught (including discouraging attitudes of faculty) more than problems with what they were taught. For a full discussion, see Chapter 22.

What Matters in College: Four Critical Years Revisited, A. Astin and G. Erlandson (eds.), Jossey-Bass, 1997.

A study of how students change and develop in college. The book shows how a range of variables—including academic programs, faculty, student peer groups, and much more—affect students' college experiences.

What Works: Building Natural Science Communities, Vols. I and II, The Independent Colleges Office, Project Kaleidoscope, Washington, DC, 1991, 1993.

These two volumes document practices in effective undergraduate programs in the natural sciences.

Department Leadership

Chairing a Mathematical Sciences Department in the 1990's, National Research Council, Washington, DC, 1990.

Proceedings of a BMS Mathematics Chairs Colloquium. Among the touchy topics covered are differential teaching loads, experimentation with new curriculum and teaching methods, and the role of applied mathematics within a mathematics department.

On Being a Department Head, John Conway, American Mathematical Society, Providence, RI, 1996.
A personalized account by a mathematics department chair at a research university.

The Academic Chairperson's Handbook, K. Beyer, N. Egly, A. Seagren, D. Wheeler, and J. Creswell, University of Nebraska Press, 1990.

A book of case studies describing successful department chairs in a variety of disciplines.

You Can Negotiate Anything, H. Cohen, Mass Market Paperback, 1989.
A text highly recommended in Doug Lind's essay (Chapter 16) for help in dealing with your dean.

Data Studies

Statistical Abstract of Undergraduate Programs in the Mathematical Sciences, Fall 1995 CBMS Survey, Mathematical Association of America, 1997.
The latest of the five-year CBMS surveys documents a surprising new trend in mathematics enrollments, namely, a 9% decline in the past five years.

A Challenge of Numbers, People in the Mathematical Sciences, B. Madison and T. Hart, eds., National Research Council, 1990.
This report is a compilation of data documenting the challenges facing the mathematical sciences community in educating students and attracting future faculty to the profession.

AMS-MAA Annual Surveys, published annually in the AMS Notices.
These reports give data about new Ph.D.'s and their employment, including starting salaries. Recent surveys have also collected data about course enrollments.

Graduate Students and Postdoctorates in Science and Engineering: Fall 1996, National Science Foundation, June 2, 1998.
This report has graduate enrollment data going back to 1966, broken down by degree program, type of institution, ethnicity, source of support, nationality, and more.

Characteristics of Recent Science and Engineering Graduates: 1995, National Science Foundation, February 28, 1998.
This report has a huge amount of information about B.S. and M.S. recipients, including continuing education data, forms of support for students, how many students hold second jobs, and more.

Characteristics of Recent Science and Engineering Graduates: 1995, National Science Foundation, February 28, 1998.
Data includes salaries and forms of employment.

Survey of Mathematics and Statistics Departments at Higher Education Institutions, National Science Foundation, December 1990.
A survey of enrollments and opinions about various problems facing mathematics and statistics departments. The sample size is considerably larger than the CBMS five-year surveys.

Undergraduate Origins of Recent (1991–95) Doctoral Recipients, National Science Foundation, April 3, 1997.
An interesting report about where recent Ph.D.'s have come from.

School Mathematics and the Training of School Mathematics Teachers

Professional Standards for Teaching Mathematics, National Council of Teachers of Mathematics, 1991.
This is the oft-cited NCTM Standards that launched the whole movement of curriculum standards in the schools. The document is intentionally vague on many details. Interpretations of what it advocates and what it downplays have been the source of considerable controversy. Note: A new set of Standards will appear in 2000.

A Call for Change: Recommendations for the Mathematical Preparation of Teachers of Mathematics, J. Leitzel, ed., Mathematical Association of America, 1991.
Recommendations for the preservice education of school mathematics teachers that are meant to prepare prospective teachers to use the NCTM Standards.

Guidelines for the Mathematical Preparation of Prospective Elementary Teachers. Texas Statewide Systemic Initiative, The Charles A. Dana Center for Mathematics and Science Education, University of Texas, Austin, 1996.

Model Standards in Mathematics for Beginning Teacher Licensing and Development: A Resource for State Dialogue, Interstate New Teacher Assessment and Support Coalition (INTASC), Council of Chief State School Officers, Washington, DC, 1995.

The Preparation of Teachers of Mathematics: Considerations and Challenges (A Letter Report), Mathematical Sciences Education Board, National Research Council, March 1996.

Towards High and Rigorous Standards for the Teaching Profession, Second Edition, National Board for Professional Teaching Standards, 1990.

What Matters Most: Teaching for America's Future, National Commission on Teaching and America's Future, 1996.

The Changing Environment in Higher Education

The following references discuss difficult issues facing higher education, particularly at research universities. These include imbalance between research and teaching, calls for a greater economic payoff from academic research, improving schools, society's changing expectations for a college education, backlash from academic "cultural wars", the view of academic researchers as just another special interest group, and more.

Academic Duty, Donald Kennedy, Harvard University Press, 1998.

An Exploration of the Nature and Quality of Undergraduate Education in Science, Mathematics, and Engineering, Sigma Xi, 1990.

Contemporary Understandings of Liberal Education: The Academy in Transition, C. Schneider and R. Shoenberg, American Association of Colleges and Universities, 1998.

Drive-Thru U., by J. Traub, The New Yorker, Oct 20 & 27, 1997, pp. 114–123.

Equilibrium in the Research University, R. Atkinson and D. Tuzin, CHANGE, May–June 1992, pp. 20–27, 30–31 (cited in Chapter 14).

From Analysis to Action: Undergraduate Education in Science, Mathematics, Engineering, and Technology, Center for Science, Mathematics, and Engineering Education, National Research Council, 1996.

Functions and Resources: The University of the Twenty-First Century, H. Shapiro, Proceedings of the University of Chicago Symposium, The University of the Twenty-First Century, 1995.

Organizing for Learning: A New Imperative, P. Ewell, American Association of Higher Education Bulletin, December 1997, pp. 3-6.

Scholarship Reconsidered: Priorities of the Professoriate, Ernest Boyer, Carnegie Foundation for the Advancement of Teaching, 1990.

Today's Academic Market Requires a New Taxonomy of Colleges, C. Finn, The Chronicle of Higher Education, 9 January 1998.

Changing the Culture: Mathematics Education in the Research Community, N. Fisher et al, eds., Conference Board of the Mathematical Sciences, Issues in Mathematics Education, Vol. 5, American Mathematical Society, 1995.

Appendices

Appendix A
AMS Task Force on Excellence

MEMBERS

Morton Lowengrub, Chair of the Task Force
Dean of the College of Arts and Sciences, Indiana University
Thomas R. Berger
Professor, Colby College
John B. Garnett
Professor, University of California, Los Angeles
Ettore Infante
Dean of the College of Arts and Sciences, Vanderbilt University
Raymond L. Johnson
Professor, University of Maryland
Barbara L. Keyfitz
Professor, University of Houston
W. James Lewis
Professor, University of Nebraska-Lincoln
Douglas Lind
Professor, University of Washington
Donald E. McClure
Professor, Brown University
Alan C. Newell
Professor, University of Arizona and University of Warwick
Alan C. Tucker
Professor, SUNY at Stony Brook
David A. Vogan, Jr.
Professor, Massachusetts Institute of Technology

AMS STAFF

Raquel E. Storti
Assistant to the Executive Director, American Mathematical Society

Chronology of the Task Force on Excellence

1992	AMS ad hoc Committee on Resource Needs for Excellence in Mathematics Instruction appointed by AMS President Michael Artin, chaired by Professor Felix Haas.
Jan 1992	Committee meets in San Antonio.
1993	Dr. Morton Lowengrub, Dean of Arts and Sciences, Indiana University assumes the chair of the Committee. The name of the Committee is changed to AMS Task Force on Excellence in Mathematics Scholarship: Assuring Quality Undergraduate and Graduate Programs at Doctoral-Granting Institutions.
May 1993	Committee meets in Chicago, IL.
Mar 1994	Task Force meets in Chicago, IL.
Aug 1994	Task Force meets in Minneapolis, MN.
	Focus Discussion I.
Oct 1994	Focus Discussion II, Washington, DC.
Jan 1995	Task Force meets in San Francisco, CA.
	Focus Discussion III, IV.
Mar 1995	Focus Discussion V, Chicago, IL.
Aug 1995	Task Force meets in Burlington, VT.
	Focus Discussion VI, VII.
Oct 1995	Focus Discussion VIII, Washington, DC.
Jan 1996	Focus Discussion IX, X, Orlando, FL.
Mar 1996	Deans Focus Discussion I, Laguna Beach, CA.
Apr 1996	Task Force meets in New York, NY.
May 1996	Deans Focus Discussion II, Chicago, IL.
Aug 1996	Focus Discussion XI, Seattle, WA.
Sep 1996	Site Visit—Oklahoma State University, Stillwater, OK.
	Site Visit—University of Michigan, Ann Arbor, MI.
Oct 1996	Site Visit—University of Chicago, Chicago, IL.
Nov 1996	Deans Focus Discussion III, Philadelphia, PA.
Dec 1996	Site Visit—University of Texas at Austin, TX.
Jan 1997	Task Force meets in San Diego, CA.
Feb 1997	Site Visit—University of Arizona, Tucson, AZ.
Apr 1997	Task Force meets in Bloomington, IN.
Oct 1998	Task Force meets in Chicago, IL.
Aug 1999	Leadership Conference, Bloomington, IN.

Appendix B
Groupings of Departments: AMS-IMS-MAA Annual Survey

(Found at http://www.ams.org/employment/groups_des.html)

The reports of the AMS-IMS-MAA Annual Survey present data for departments divided into groups according to several characteristics, the principal one being the highest degree offered in the mathematical sciences. Doctoral-granting departments of mathematics are further subgrouped according to their ranking by "scholarly quality of program faculty", as reported in the 1995 publication *Research-Doctorate Programs in the United States: Continuity and Change.*[1] These rankings update those reported previously in a study published in 1982.[2] Consequently, the departments that now (in 1996) comprise Groups I, II, and III differ from those used in prior surveys. These groupings are used for statistical reporting purposes only and may not accurately reflect current program quality at individual departments.

The subdivision of the Group I institutions into Group I Public and Group I Private is new with the 1996 Annual Survey. With the increase in the size of the Group I departments from 39 to 48, the AMS-IMS-MAA Data Committee judged that a further subdivision along the lines of public and private would provide more meaningful reporting of the data for these departments.

---

[1] *Research-Doctorate Programs in the United States: Continuity and Change*, edited by Marvin L. Goldberger, Brendan A. Maher, and Pamela Ebert Flattau; National Academy Press, Washington, D.C., 1995.

[2] *An Assessment of Research-Doctorate Programs in the United States: Mathematical and Physical Sciences*, edited by Lyle V. Jones, Gardner Lindzey, and Porter E. Coggeshall; National Academy Press, Washington, D.C., 1982. The information on mathematics, statistics, and computer science was presented in digest form in the April 1983 issue of the *Notices*, pages 257–267, and an analysis of the classifications was given in the June 1983 *Notices*, pages 392–393.

Brief descriptions of all the groupings are as follows:

- **Group I** is composed of 48 departments with scores in the 3.00–5.00 range.
- **Group I Public** and **Group I Private** are Group I departments at public institutions and private institutions, respectively.
- **Group II** is composed of 56 departments with scores in the 2.00–2.99 range.
- **Group III** contains the remaining U.S. departments reporting a doctoral program, including a number of departments not included in the 1995 ranking of program faculty.
- **Group IV** contains U.S. departments (or programs) of statistics, biostatistics, and biometrics reporting a doctoral program.
- **Group V** contains U.S. departments (or programs) in applied mathematics/applied science, operations research, and management science which report a doctoral program.
- **Group Va** is applied mathematics/applied science; **Group Vb** is operations research and management science.
- **Group M** contains U.S. departments granting a master's degree as the highest graduate degree.
- **Group B** contains U.S. departments granting a baccalaureate degree only.

Group I Public

(Scores 3.00–5.00: 25 departments)

City University of New York, Graduate Center
Georgia Institute of Technology
Indiana University, Bloomington
Michigan State University
Ohio State University
Pennsylvania State University
Purdue University
Rutgers University, New Brunswick
State University of New York, Stony Brook
University of California, Berkeley
University of California, Los Angeles
University of California, San Diego
University of California, Santa Barbara
University of Illinois, Chicago
University of Illinois, Urbana-Champaign
University of Maryland, College Park
University of Michigan, Ann Arbor
University of Minnesota, Minneapolis
University of North Carolina, Chapel Hill
University of Oregon
University of Texas, Austin
University of Utah
University of Virginia
University of Washington
University of Wisconsin, Madison

Group I Private

(Scores 3.00–5.00: 23 departments)

Boston University
Brandeis University
Brown University
California Institute of Technology
Carnegie Mellon University
Columbia University
Cornell University
Duke University
Harvard University
Johns Hopkins University
Massachusetts Institute of Technology
New York University, Courant Institute

Northwestern University
Princeton University
Rensselaer Polytechic Institute
Rice University
Stanford University
University of Chicago
University of Notre Dame
University of Pennsylvania
University of Southern California
Washington University
Yale University

Group II

(Scores 2.00–2.99: 56 departments)

Arizona State University
Auburn University
Case Western Reserve University
Claremont Graduate University
Clemson University
Colorado State University
Dartmouth College
Florida State University
Iowa State University
Kansas State University
Kent State University
Lehigh University
Louisiana State University, Baton Rouge
North Carolina State University, Raleigh
Northeastern University
Oregon State University
Polytechnic University
State University of New York, Albany
State University of New York, Binghamton
State University of New York, Buffalo
Syracuse University
Temple University
Texas A&M University
Texas Tech University
Tulane University
University of Arizona
University of California, Davis
University of California, Irvine
University of California, Riverside
University of California, Santa Cruz
University of Cincinnati

University of Colorado, Boulder
University of Connecticut
University of Delaware
University of Florida
University of Georgia
University of Hawaii
University of Houston
University of Iowa
University of Kentucky
University of Massachusetts, Amherst
University of Miami
University of Missouri, Columbia
University of Nebraska, Lincoln
University of North Texas
University of Oklahoma
University of Pittsburgh
University of Rochester
University of South Carolina
University of Tennessee
University of Texas, Arlington
Vanderbilt University
Virginia Polytechnic Institute & State University
Washington State University
Wayne State University
Wesleyan University

Group III

(Scores below 2.00: 29 departments)

Adelphi University
Bowling Green State University
Clarkson University
Colorado School of Mines
Drexel University
George Washington University
Howard University
Idaho State University
Illinois State University
New Mexico State University
Northern Illinois University
Ohio University
Old Dominion University
Southern Illinois University, Carbondale
Southern Methodist University
St. Louis University
Stevens Institute of Technology

University of Alabama, Huntsville
University of Alabama, Tuscaloosa
University of Maryland, Baltimore
University of Mississippi
University of Missouri, Rolla
University of Rhode Island
University of South Florida
University of Southwestern Louisiana
University of Texas, Dallas
University of Wisconsin, Milwaukee
University of Wyoming
Western Michigan University

(Not included in the 1995 NRC study: 43 departments)

Air Force Institute of Technology
American University
Brigham Young University
Bryn Mawr College
Catholic University of America
Central Michigan University
Clark University
College of William & Mary
Emory University
Florida Atlantic University
Indiana University-Purdue University
Marquette University
Michigan Technological University
Mississippi State University
Montana State University
Naval Postgraduate School
New Jersey Institute of Technology
North Dakota State University
Oklahoma State University
Portland State University
Rutgers University, Newark
Tufts University
University of Alabama, Birmingham
University of Alaska, Fairbanks
University of Arkansas
University of Central Florida
University of Colorado, Denver
University of Denver
University of Idaho
University of Kansas *
University of Memphis

University of Missouri, Kansas City
University of Montana
University of New Hampshire
University of New Mexico *
University of North Carolina, Charlotte
University of Northern Colorado
University of Toledo
University of Vermont
Utah State University
West Virginia University
Wichita State University
Worcester Polytechnic Institute

* These departments were formerly in Group II based on the 1982 NRC rankings.

Group IV

(Statistics, biostatistics, and biometrics: 81 departments)

Auburn University, Discrete & Statistical Sciences
Carnegie Mellon University, Statistics
Case Western Reserve University, Statistics
Case Western Reserve University, Epidemiology & Biostatistics
Colorado State University, Statistics
Columbia University, Statistics
Columbia University, Biostatistics
Cornell University, Statistics
Cornell University, Biometrics
Cornell University, Social Statistics
Duke University, Statistics & Decision Sciences
Emory University, Biostatistics
Florida State University, Statistics
George Mason University, Applied & Engineering Statistics
George Washington University, Statistics
Harvard University, Statistics
Harvard University, Biostatistics
Iowa State University, Statistics
Johns Hopkins University, Biostatistics
Kansas State University, Statistics
Massachusetts Institute of Technology, Statistics
Medical University of South Carolina, Biometry & Epidemiology
Michigan State University, Statistics & Probability
New York University, Statistics & Operations Research
North Carolina State University, Raleigh, Statistics
North Dakota State University, Statistics
Northwestern University, Statistics
Ohio State University, Statistics

Oklahoma State University, Statistics
Oregon State University, Statistics
Pennsylvania State University, Statistics
Purdue University, Statistics
Rice University, Statistics
Rutgers University, New Brunswick, Statistics
Southern Methodist University, Statistical Science
Stanford University, Statistics
State University of New York, Albany, Statistics & Biometry
State University of New York, Buffalo, Statistics
Temple University, Statistics
Texas A&M University, Statistics
University of Alabama, Birmingham, Biostatistics
University of Alabama, Tuscaloosa, Applied Statistics
University of California, Berkeley, Statistics
University of California, Berkeley, Biostatistics
University of California, Davis, Statistics
University of California, Los Angeles, Biostatistics
University of California, Riverside, Statistics
University of California, Santa Barbara, Statistics & Applied Probability
University of Chicago, Statistics
University of Cincinnati, Epidemiology & Biostatistics, Medical College
University of Connecticut, Statistics
University of Florida, Statistics
University of Georgia, Statistics
University of Hawaii, Public Health Sciences
University of Illinois, Urbana-Champaign, Statistics
University of Iowa, Statistics & Actuarial Science
University of Kentucky, Statistics
University of Maryland, College Park, Measure Statistics
University of Michigan, Ann Arbor, Statistics
University of Michigan, Ann Arbor, Biostatistics
University of Minnesota, Minneapolis, Statistics
University of Minnesota, Minneapolis, Biostatistics
University of Missouri, Columbia, Statistics
University of North Carolina, Chapel Hill, Statistics
University of North Carolina, Chapel Hill, Biostatistics
University of Oklahoma, Biostatistics & Epidemiology
University of Pennsylvania, Statistics
University of Pittsburgh, Statistics
University of Pittsburgh, Biostatistics
University of Rochester, Statistics
University of South Carolina, Statistics
University of Virginia, Statistics
University of Washington, Statistics
University of Washington, Biostatistics

University of Wisconsin, Madison, Statistics
University of Wyoming, Statistics
Virginia Commonwealth University, Biostatistics
Virginia Polytechnic Institute & State University, Statistics
West Virginia University, Statistics & Computer Science
Yale University, Statistics
Yale University, Biostatistics

Group Va

(Applied mathematics/ applied science: 18 departments)

Brown University, Applied Mathematics
California Institute of Technology, Applied Mathematics
Cornell University, Applied Mathematics
Florida Institute of Technology, Applied Mathematics
Harvard University, Engineering & Applied Sciences
Johns Hopkins University, Mathematical Sciences
Northwestern University, Engineering Science & Applied Mathematics
Princeton University, Applied & Computational Mathematics
Rice University, Computational & Applied Mathematics
State University of New York, Stony Brook, Applied Mathematics & Statistics
University of Arizona, Applied Mathematics
University of Colorado, Boulder, Applied Mathematics
University of Iowa, Applied Mathematical & Computational Sciences
University of Louisville, Engineering Mathematics & Computer Science
University of Texas, Austin, Computational & Applied Mathematics
University of Virginia, Applied Mathematics & Mechanics
University of Washington, Applied Mathematics
Washington University, Systems Science & Mathematics

Group Vb

(Operations research and management science: 31 departments)

Case Western Reserve University, Operations Research
Cornell University, Operations Research & Industrial Engineering
George Mason University, Operations Research & Engineering
George Washington University, Operations Research
Georgia Institute of Technology, Industrial & Systems Engineering
Massachusetts Institute of Technology, Operations Research
Massachusetts Institute of Technology, Management Science
Naval Postgraduate School, Operations Research
North Carolina State University, Raleigh, Operations Research
Northwestern University, Managerial Economics & Decision Science
Northwestern University, Industrial Engineering & Management Science
Purdue University, Industrial Engineering

Rensselaer Polytechic Institute, Decision Science & Engineering Systems
Rutgers University, New Brunswick, Operations Research
Stanford University, Engineering-Economic Systems & Operations Research
State University of New York, Buffalo, Industrial Engineering
Syracuse University, Industrial Engineering & Operations Research
Union College, Administrative & Engineering Systems
University of Alabama, Tuscaloosa, Management Science & Statistics
University of California, Berkeley, Industrial Engineering & Op Research
University of Chicago, Graduate School of Business
University of Cincinnati, Quantitative Analysis & Operations Management
University of Florida, Industrial & Systems Engineering
University of Miami, Management Science
University of Michigan, Ann Arbor, Industrial & Operations Engineering
University of Minnesota, Minneapolis, Management Science
University of North Carolina, Chapel Hill, Operations Research
University of Pittsburgh, Industrial Engineering
University of Tennessee, Management Science
University of Wisconsin, Madison, Industrial Engineering
Virginia Polytechnic Institute & State University, Indus & Systems Engineering

Appendix C
The Carnegie Foundation Classification of Higher Education –

(Found at http://www.carnegiefoundation.org/cihe/)

Foreword (excerpts)

Ernest L . Boyer

The Carnegie Classification of higher education groups American colleges and universities according to their missions. This classification was developed by Clark Kerr in 1970 primarily to improve the precision of the Carnegie Commission's research. Over the years, the system has gained credibility and served as a helpful guide for scholars and researchers.

The Carnegie Classification is not intended to establish a hierarchy among higher learning institutions. Rather, the aim is to cluster institutions with similar programs and purposes, and we oppose the use of the classification as a way of making qualitative distinctions among the separate sectors. We have, in this country, a rich array of institutions serving a variety of needs, and there are institutions of distinction in every category of the Carnegie Classification.

Over the years, we have modified the definitions somewhat to improve the groupings in this new edition, the most consequential change we've made is to classify all institutions, for the first time, according to the highest level of degree conferred—from associate of arts to doctoral degrees. This means that the "Liberal Arts" category—which will now be called "Baccalaureate"—includes all colleges where the baccalaureate is the highest degree awarded. The "Comprehensive" category—which will now be called "Master's (Comprehensive)" includes master's–granting institutions. We're convinced that classifying campuses on the basis of degree level brings still more clarity and objectivity to the process.

Looking for larger patterns we are once again impressed that with all the talk about cutbacks and retrenchment over 400 new institutions appear in this edition —the majority being two-year institutions listed in the Associate of Arts category. Approximately 100 of the new colleges are specialized institutions. This growth is counterbalanced by over 200 institutions that merged, closed, or otherwise are no longer eligible for inclusion in this listing. The overall number of institutions in the 1994 Carnegie Classification increased from 3,389 to 3,595. The new Carnegie Classification also reveals what some have called the "upward drift" in higher education, and of special interest is the continuing expansion of research and doctoral institutions. America must continue to support a core of world-class research centers; they are essential to the advancement of knowledge and to human achievement. Such activity is costly, however, and it is crucial that we have available the fiscal resources needed to sustain an expanding network of institutions devoted to scholarly research.

We also note, with satisfaction that the balance between the private and public sector has, since 1987 remained relatively constant and, in spite of earlier trends and dark predictions, the independent colleges in America have shown resiliency and growth. We urge that public policy continue to acknowledge the contributions of both sectors.

...

In summary, the 1994 Carnegie Classification reveals a healthy and expanding network of higher learning institutions in the nation. Voices of gloom and predictions of decline are not supported by the trends. Americans, perhaps as never before need a vibrant system of higher education one that is closely tied to the economic and social vitality of the nation and to the private hopes of students and their families

Colleges and universities in the United States have an amazing capacity to respond creatively to new conditions. This system, accomplished without a "master plan" and federal directive remains one of America's most remarkable achievements.

DEFINITIONS OF CATEGORIES

The 1994 Carnegie Classification includes all colleges and universities in the United States that are degree-granting and accredited by an agency recognized by the U.S. Secretary of Education.

Research Universities I: These institutions offer a full range of baccalaureate programs, are committed to graduate education through the doctorate, and give high priority to research. They award 50 or more doctoral degrees[1] each year. In addition, they receive annually $40 million or more in federal support.[2]

Research Universities II: These institutions offer a full range of baccalaureate programs, are committed to graduate education through the doctorate, and give high priority to research. They award 50 or more doctoral degrees[1] each year. In addition, they receive annually between $15.5 million and $40 million in federal support.[2]

Doctoral Universities I: These institutions offer a full range of baccalaureate programs and are committed to graduate education through the doctorate. They award at least 40 doctoral degrees[1] annually in five or more disciplines.[3]

Doctoral Universities II: These institutions offer a full range of baccalaureate programs and are committed to graduate education through the doctorate. They award annually at least ten doctoral degrees in three or more disciplines, or 20 or more doctoral degrees in one or more disciplines.[3]

Master's (Comprehensive) Universities and Colleges I: These institutions offer a full range of baccalaureate programs and are committed to graduate education through the master's degree. They award 40 or more master's degrees annually in three or more disciplines.[3]

Master's (Comprehensive) Universities and Colleges II: These institutions offer a full range of baccalaureate programs and are committed to graduate education through the master's degree. They award 20 or more master's degrees annually in one or more disciplines.[3]

Baccalaureate (Liberal Arts) Colleges I: These institutions are primarily undergraduate colleges with major emphasis on baccalaureate-degree programs. They award 40 percent or more of their baccalaureate degrees in liberal arts fields[4] and are restrictive in admissions.

Baccalaureate Colleges II: These institutions are primarily undergraduate colleges with major emphasis on baccalaureate-degree programs. They award less than 40 percent of their baccalaureate degrees in liberal arts fields[4] or are less restrictive in admissions.

Associate of Arts Colleges: These institutions offer associate of arts certificate or degree programs and, with few exceptions, offer no baccalaureate degrees.[5]

Specialized Institutions: These institutions offer degrees ranging from the bachelor's to the doctorate. At least 50 percent of the degrees awarded by these institutions are in a single discipline. Specialized institutions include: theological seminaries, bible colleges, medical schools, schools of engineering and technology, schools of business and management, schools of art and design, schools of music, schools of law, teachers' colleges, graduate centers, maritime academies, military institutes, and tribal colleges.

Notes on Definitions

[1]Doctoral degrees include Doctor of Education, Doctor of Juridical Science, Doctor of Public Health, and the Ph.D. in any field.

[2]Total federal obligation figures are available from the National Science Foundation's annual report called "Federal Support to Universities, Colleges, and Nonprofit Institutions". The years used in averaging total federal obligations are 1989, 1990, and 1991.

[3]Distinct disciplines are determined by the U.S. Department of Education's Classification of Instructional Programs' 4-digit series.

[4]The liberal arts disciplines include English language and literature, foreign languages, letters, liberal and general studies, life sciences, mathematics, philosophy and religion, physical sciences, psychology, social sciences, the visual and performing arts, area and ethnic studies, and multi- and interdisciplinary studies. The occupational and technical disciplines include agriculture, allied health, architecture, business and management, communications, conservation and natural resources, education, engineering, health sciences, home economics, law and legal studies, library and archival sciences, marketing and distribution, military sciences, protective services, public administration andservices, and theology.

[5]This group includes community, junior, and technical colleges.

Research and Doctoral Universities

Research Universities I (Public)

ALABAMA
University of Alabama at Birmingham
ARIZONA
Arizona State University
University of Arizona
CALIFORNIA
University of California at Berkeley
University of California at Davis
University of California at Irvine
University of California at Los Angeles
University of California at San Diego
University of California at San Francisco
University of California at Santa Barbara
COLORADO
Colorado State University
University of Colorado at Boulder
CONNECTICUT
University of Connecticut
FLORIDA
Florida State University
University of Florida
GEORGIA
Georgia Institute of Technology
University of Georgia
HAWAII
University of Hawaii at Manoa
ILLINOIS
University of Illinois at Chicago
University of Illinois at Champaign-Urbana
INDIANA
Indiana University at Bloomington
Purdue University, Main Campus
IOWA
Iowa State University
University of Iowa
KANSAS
University of Kansas, Main Campus
KENTUCKY
University of Kentucky
LOUISIANA
Louisiana State University and Agricultural and Mechanical College
MARYLAND
University of Maryland at College Park
MASSACHUSETTS
University of Massachusetts at Amherst
MICHIGAN
Michigan State University
University of Michigan at Ann Arbor
Wayne State University
MINNESOTA
University of Minnesota at Twin Cities
MISSOURI
University of Missouri at Columbia
NEBRASKA
University of Nebraska at Lincoln
NEW JERSEY
Rutgers, the State University of New Jersey, New Brunswick Campus
NEW MEXICO
New Mexico State University, Main Campus
University of New Mexico, Main Campus
NEW YORK
State University of New York at Buffalo
State University of New York at Stony Brook
NORTH CAROLINA
North Carolina State University
University of North Carolina at Chapel Hill
OHIO
Ohio State University, Main Campus, The
University of Cincinnati, Main Campus
OREGON
Oregon State University
PENNSYLVANIA
Pennsylvania State University, Main Campus
Temple University
University of Pittsburgh, Pittsburgh Campus
TENNESSEE
University of Tennessee at Knoxville
TEXAS
Texas A&M University
University of Texas at Austin
UTAH
University of Utah
Utah State University
VIRGINIA
University of Virginia
Virginia Commonwealth University
Virginia Polytechnic Institute and State University
WASHINGTON
University of Washington
WEST VIRGINIA
West Virginia University
WISCONSIN
University of Wisconsin at Madison

Research Universities I (Private)

CALIFORNIA
California Institute of Technology
Stanford University
University of Southern California
CONNECTICUT
Yale University
DISTRICT OF COLUMBIA
Georgetown University
Howard University
FLORIDA
University of Miami
GEORGIA
Emory University
ILLINOIS
Northwestern University
University of Chicago
MARYLAND
Johns Hopkins University
MASSACHUSETTS
Boston University
Harvard University
Massachusetts Institute of Technology
Tufts University
MISSOURI
Washington University
NEW JERSEY
Princeton University
NEW YORK
Columbia University in the City of New York
Cornell University
New York University
Rockefeller University
University of Rochester
Yeshiva University
NORTH CAROLINA
Duke University
OHIO
Case Western Reserve University
PENNSYLVANIA
Carnegie Mellon University
University of Pennsylvania
RHODE ISLAND
Brown University
TENNESSEE
Vanderbilt University

Research Universities II (Public)

ALABAMA
Auburn University
ARKANSAS
University of Arkansas, Main Campus
CALIFORNIA
University of California at Riverside
University of California at Santa Cruz
DELAWARE
University of Delaware
FLORIDA
University of South Florida
IDAHO
University of Idaho
ILLINOIS
Southern Illinois University at Carbondale
KANSAS
Kansas State University
MISSISSIPPI
Mississippi State University
University of Mississippi
NEW YORK
State University of New York at Albany
OHIO
Kent State University, Main Campus
Ohio University, Main Campus
OKLAHOMA
Oklahoma State University, Main Campus
University of Oklahoma, Norman Campus
OREGON
University of Oregon
RHODE ISLAND
University of Rhode Island
SOUTH CAROLINA
Clemson University
University of South Carolina at Columbia
TEXAS
Texas Tech University
University of Houston
VERMONT
University of Vermont
WASHINGTON
Washington State University
WISCONSIN
University of Wisconsin at Milwaukee
WYOMING
University of Wyoming

Research Universities II (Private)

DISTRICT OF COLUMBIA
George Washington University
INDIANA
University of Notre Dame
LOUISIANA
Tulane University
MASSACHUSETTS
Brandeis University
Northeastern University
MISSOURI
Saint Louis University
NEW YORK
Rensselaer Polytechnic Institute
Syracuse University, Main Campus
PENNSYLVANIA
Lehigh University
TEXAS
Rice University
UTAH
Brigham Young University

Doctoral Universities I (Public)

ALABAMA
University of Alabama, The
ARIZONA
Northern Arizona University
COLORADO
University of Northern Colorado
GEORGIA
Georgia State University
ILLINOIS
Illinois State University
Northern Illinois University
INDIANA
Ball State University
KENTUCKY
University of Louisville
MICHIGAN
Western Michigan University
MISSISSIPPI
University of Southern Mississippi
MISSOURI
University of Missouri at Kansas City
University of Missouri at Rolla
NEW YORK
City University of New York Graduate School and University Center
State University of New York at Binghamton
NORTH CAROLINA
University of North Carolina at Greensboro
OHIO
Bowling Green State University
Miami University
University of Akron, Main Campus
University of Toledo
PENNSYLVANIA
Indiana University of Pennsylvania
TENNESSEE
Memphis State University
TEXAS
East Texas State University
Texas Woman's University
University of North Texas
University of Texas at Arlington
University of Texas at Dallas
VIRGINIA
College of William and Mary
Old Dominion University

Doctoral Universities I (Private)

CALIFORNIA
Claremont Graduate School
United States International University
COLORADO
University of Denver
DISTRICT OF COLUMBIA
American University, The
Catholic University of America
FLORIDA
Florida Institute of Technology
Nova University
GEORGIA
Clark Atlanta University
ILLINOIS
Illinois Institute of Technology
Loyola University of Chicago
MASSACHUSETTS
Boston College
MICHIGAN
Andrews University
NEW YORK
Adelphi University
Fordham University

Hofstra University
New School for Social Research
Polytechnic University
Saint John's University
Teachers College, Columbia University
OHIO
Union Institute
PENNSYLVANIA
Drexel University
TEXAS
Southern Methodist University
WISCONSIN
Marquette University

Doctoral Universities II (Public)

ALABAMA
University of Alabama at Huntsville
ALASKA
University of Alaska at Fairbanks
CALIFORNIA
San Diego State University
COLORADO
Colorado School of Mines
University of Colorado at Denver
FLORIDA
Florida Atlantic University
Florida International University
University of Central Florida
IDAHO
Idaho State University
INDIANA
Indiana State University
Indiana University-Purdue University at Indianapolis
KANSAS
Wichita State University, The
LOUISIANA
Louisiana Tech University
University of New Orleans
University of Southwestern Louisiana
MAINE
University of Maine
MARYLAND
University of Maryland Baltimore County
MASSACHUSETTS
University of Massachusetts at Lowell
MICHIGAN
Michigan Technological University
MISSOURI
University of Missouri at Saint Louis
MONTANA
Montana State University
University of Montana, The
NEVADA
University of Nevada, Reno
NEW HAMPSHIRE
University of New Hampshire
NEW JERSEY
New Jersey Institute of Technology
Rutgers, The State University of New Jersey, Newark Campus
NEW YORK
State University of New York College of Environmental Science and Forestry
NORTH DAKOTA
North Dakota State University, Main Campus
University of North Dakota, Main Campus
OHIO
Cleveland State University
Wright State University, Main Campus
OREGON
Portland State University
SOUTH DAKOTA
University of South Dakota
TENNESSEE
Middle Tennessee State University
Tennessee State University
TEXAS
Texas Southern University
VIRGINIA
George Mason University
PUERTO RICO
University of Puerto Rico, Rio Piedras Campus

Doctoral Universities II (Private)

CALIFORNIA
Biola University
Loma Linda University
Pepperdine University
University of LaVerne
University of San Diego
University of San Francisco
University of the Pacific
ILLINOIS
De Paul University
MASSACHUSETTS
Clark University
Worcester Polytechnic Institute
MICHIGAN
University of Detroit, Mercy
NEW HAMPSHIRE
Dartmouth College

NEW JERSEY
Seton Hall University
Stevens Institute of Technology
NEW YORK
Clarkson University
Pace University
NORTH CAROLINA
Wake Forest University
OKLAHOMA
University of Tulsa
PENNSYLVANIA
Duquesne University
Hahnemann University
TEXAS
Baylor University
Texas Christian University

Appendix D
National Science Foundation Programs

A comprehensive and up-to-date list of NSF programs can be found at:
http://www.nsf.gov/home/programs/start.htm.
Excerpts from the NSF Web site are included below to illustrate the kinds of information available about specific divisions or programs. Navigating the Web site also provides an overall view of the structure of the National Science Foundation—helpful knowledge when dealing with your administration or the Foundation itself.

Division of Mathematical Sciences (DMS)

The Division of Mathematical Sciences (DMS) supports a wide range of projects aimed at developing and exploring the properties and applications of mathematical structures. Most of these projects are those awarded to single investigators or small groups of investigators working with graduate students and postdoctoral researchers. Programs such as Mathematical Sciences Infrastructure handle activities that fall outside this mode.

DMS supports research through the following programs and activities:

- Algebra And Number Theory
- Applied Mathematics
- Analysis
- Computational Mathematics
- Geometric Analysis
- Statistics And Probability
- Topology And Foundations
- Mathematical Sciences Infrastructure Program
- Grants For Vertical Integration Of Research And Education
- Cross-Disciplinary Interactions

Proposals submitted to DMS for general conferences, workshops, symposia, special years, and related activities should be submitted to the appropriate disciplinary program. Proposals should be submitted one year in advance of the start of the activity. Contact the Division for information on proposal requirements.

In addition to the usual types of research grants awarded to principal investigators and institutions, DMS supports the following:

- **University/Industry Cooperative Research.** DMS feels it is important to provide more opportunities to conduct research and training in an industrial environment and for industrial scientists to return periodically to academia. To facilitate both research and training, the Division provides Mathematical Sciences University/Industry Postdoctoral Research Fellowships, Senior Research Fellowships, and Industry-Based Graduate Research Assistantships and Cooperative Fellowships in the Mathematical Sciences.
- **Interdisciplinary Grants**. These grants enable faculty to expand their skills and knowledge into areas beyond their disciplinary expertise, and to subsequently apply the knowledge to their research as well as enrich the educational experiences and career options for students. These grants support interdisciplinary experiences at the principal investigator's (PI's) institution (outside of the PI's department), or at different institutions such as academic, financial, and industrial institutions, in a nonmathematical science environment.

Sample Programs:

- Mid-Career Methodological Opportunities (NSF 99-33)
- Integrative Graduate Education and Research Training Program (IGERT)(NSF 98-96)
- Optimized Portable Algorithms and Application Libraries (OPAAL) (NSF 98-64)
- Knowledge and Distributed Intelligence (NSF 99-29)
- Scientific Computing Research Environments in the Mathematical Sciences (NSF 99-48)
- Grants for Vertical Integration of Research and Education in the Mathematical Sciences (VIGRE) (NSF 99-16)
- Professional Opportunities for Women in Research and Education (POWRE) (NSF 98-160)
- Grant Opportunities for Academic Liaison with Industry (GOALI) (NSF 98-142)
- Mathematical Sciences Postdoctoral Research Fellowships (NSF 98-135)
- Faculty Early Career Development (CAREER) Program (NSF 98-103)
- Interdisciplinary Grants in the Mathematical Sciences (NSF 98-145)

Education and Human Resources (EHR)

The Directorate for Education and Human Resources (EHR) has primary responsibility for NSF's efforts to provide national leadership in improving science, mathematics, engineering, and technology education. Its comprehensive and coordinated programs address every education level (i.e., pre-kindergarten through postdoctoral study), as well as early career development and science literacy in the general public.

EHR has five major long-term goals that provide the focus for the various activities of the seven divisions/offices described here. These goals ensure that:

- Standards-based science and mathematics education is available to every child in the United States, thus enabling all who have interest and talent to pursue technical careers at all levels;
- The educational pipelines that carry students to careers in science, mathematics, and engineering yield adequate numbers of well-educated individuals who can meet the needs of the technical workplace in the United States;
- Those who select science or engineering careers have available the best professional undergraduate and graduate education, and opportunities are available at the college level for interested nonspecialists to broaden their scientific backgrounds;
- The instructional workforce has the disciplinary and pedagogical skills to ensure an excellent education for every student and learner; and
- Opportunities for quality informal science education are available to maintain public interest in, and awareness of, scientific and technological developments.

EHR programs intend to reform education venues and strengthen education pipelines, so that all students are well prepared for an increasingly technology-driven society and workplace. Programmatic foci of the directorate include systemic reform of science and mathematics education in grades K–12, and the development of resources critical to that reform; preparation of the instructional workforce; achievement of an integrated understanding of institutional reform at the undergraduate level; cultivating a research base of knowledge for implementing innovative reform strategies in grades K–16; advanced training of scientists, mathematicians, and engineers for the 21st century; and the application of technology across all education levels (of particular interest are projects that integrate content, technology, and pedagogy).

The EHR Directorate comprises the following Divisions:

- Division of Educational System Reform (ESR)
- Division of Elementary, Secondary, and Informal Education (ESIE)
- Division of Undergraduate Education (DUE)
- Division of Graduate Education (DGE)
- Division of Human Resource Development (HRD)
- Division of Research, Evaluation, and Communication (REC)
- Experimental Program to Stimulate Competitive Research (EPSCoR)

Division of Undergraduate Education (DUE)

Within EHR the Division of Undergraduate Education (DUE) serves as the focal point for NSF's efforts in undergraduate education. Whether preparing students to participate as citizens in a technological society, to enter the work force with two- or four-year degrees, to continue their formal education in graduate school, or to further their education in response to new career goals or workplace expectations, undergraduate education provides the critical link between the Nation's secondary schools and a society increasingly dependent on science and technology.

DUE's programs and leadership efforts aim to strengthen the vitality of undergraduate science, mathematics, engineering, and technology (SMET) education for all students, including SMET majors, prospective teachers of grades pre-K–12, students preparing for the technical workplace, and students in their role as citizens in a technological society.

Projects submitted to programs in DUE are encouraged to incorporate, as appropriate, features that address one or more of four themes that have been targeted for special emphasis. These themes are teacher preparation, professional development for faculty, increasing diversity within SMET fields, and integrating technology in education. Although the activities described below are expected to constitute the majority of projects supported through DUE, proposals that address other mechanisms for improving undergraduate SMET education will be considered.

DUE supports research through the following programs and activities:

- Advanced Technological Education
- Course, Curriculum, and Laboratory Improvement
- NSF Collaboratives for Excellence in Teacher Preparation

Sample Programs:

- Advanced Technological Education (NSF 99-53)
- Centers of Research Excellence in Science and Technology (CREST)
- Collaborative Research on Learning Technologies (CRLT)
- Course, Curriculum, and Laboratory Improvement (NSF 99-53)
- Graduate Teaching Fellows in K–12 Education (TBA)
- Integrative Graduate Education and Research Training Program (IGERT) (NSF98-96)
- Optimized Portable Algorithms and Application Libraries (OPAAL) (NSF 98-64)
- Professional Opportunities for Women in Research and Education (POWRE) (NSF 98-160)
- Major Research Instrumentation Program (NSF98-16)
- Minority Research Planning Grants and Career Advancement
- New Computational Challenges (NCC)
- NSF Collaboratives for Excellence in Teacher Preparation (NSF 99-53)
- Presidential Early Career Awards for Scientists and Engineers
- Research Experiences for Undergraduates
- Research in Undergraduate Institutions

- Research Opportunity Awards
- Urban Research Initiative